Springer Tracts on Transportation and Traffic

Volume 13

Series editor

Roger P. Roess, New York University Polytechnic School of Engineering,
New York, USA
e-mail: rpr246@nyu.edu

About this Series

The book series "Springer Tracts on Transportation and Traffic" (STTT) publishes current and historical insights and new developments in the fields of Transportation and Traffic research. The intent is to cover all the technical contents, applications, and multidisciplinary aspects of Transportation and Traffic, as well as the methodologies behind them. The objective of the book series is to publish monographs, handbooks, selected contributions from specialized conferences and workshops, and textbooks, rapidly and informally but with a high quality. The STTT book series is intended to cover both the state-of-the-art and recent developments, hence leading to deeper insight and understanding in Transportation and Traffic Engineering. The series provides valuable references for researchers, engineering practitioners, graduate students and communicates new findings to a large interdisciplinary audience.

More information about this series at http://www.springer.com/series/11059

Hyung-Suk Han · Dong-Sung Kim

Magnetic Levitation

Maglev Technology and Applications

 Springer

Hyung-Suk Han
Department of Magnetic Levitation
Korea Institute of Machinery and Materials
Daejeon
Republic of Korea

Dong-Sung Kim
Department of Magnetic Levitation
Korea Institute of Machinery and Materials
Daejeon
Republic of Korea

ISSN 2194-8119   ISSN 2194-8127   (electronic)
Springer Tracts on Transportation and Traffic
ISBN 978-94-017-7522-9   ISBN 978-94-017-7524-3   (eBook)
DOI 10.1007/978-94-017-7524-3

Library of Congress Control Number: 2015958901

Printed on acid-free paper

This Springer imprint is published by SpringerNature
The registered company is Springer Science+Business Media B.V. Dordrecht

Foreword

In 2004 and 2005, high- and low-speed magnetic trains developed based on levitation technology was commercialized for a public transportation in China and Japan, respectively. For the last decade, they demonstrated their outstanding performance in various aspects of environmental friendliness, reliability, safety, and low operation and maintenance cost. Owing to this success, magnetically levitated trains will become a promising transportation option in the future, in combination with conventional wheel-on-rail systems. Moreover, the applications of magnetic levitation will be increasing to high-speed transfer systems such as super elevators. Therefore, this is the opportune moment to have this definitive and comprehensive work on the topic.

The medium for levitation is the magnetic fields produced by permanent and superconducting magnets and electromagnets in a static or dynamic mode. The interaction between the magnetic fields produces either repulsive or attractive forces, which permit the levitation and propulsion of the objects without physical contact. A higher lift force/magnet weight ratio is desirable for efficient energy consumption. In addition, sufficient damping is required to guarantee smoother operation. Despite the advantage of requiring no energy, the application of permanent magnets in levitation has been rather limited because of their low lift force/magnet weight ratio and damping. Owing to the discovery of new magnetic materials and the invention of the Halbach array enabling an increased lift force/magnet weight ratio, however, permanent magnets are attracting more attention today. It is believed that the Halbach array, in particular, has much more potential. Superconducting magnets with a very high-strength magnetic field, i.e., higher lift force/magnet weight ratio, have been under development for many years. It resulted in the construction of a high-speed railway system L0, which will be in service in near future. Although this magnet requires the cryostat to maintain its superconductivity with low damping, it might be more attractive when high-temperature superconductors are discovered. The most widely used electromagnets inevitably need sophisticated feedback control systems to maintain constant separation between the objects. Once stable levitation control is available, it might be more attractive due to its capacity to adjust its motion with high precision and damping. In conclusion, the recent advances in power electronics and sensing technologies open up new applications of electromagnetic levitation that can provide lightweight systems and high-precision motion control.

The development of magnetic materials and electronic devices could offer more opportunities for us than ever before, including new magnetically levitated applications in railway, automation, aerospace, and entertainment. In particular, personal rapid transit could be realized by employing a suitable levitation technology for a particular application. Levitation technology, once a staple of science fiction movies and novels, will become a disruptive technology in the near future.

Yong-Taek Im
President of Korea Institute of Machinery and Materials

Preface

Magnetic levitation (suspension) for contactless operation has been in development as an alternative to wheel-on-rail systems since Graeminger first patented an electromagnetic suspension device in 1912. This led to a number of operational and experimental magnetic trains being constructed, for both low- and high-speed operations, which are currently in service or will be in several years. The well-known Transrapid, which runs at a maximum operating speed of 430 km/h, has been successfully operating without any operational problems since it was unveiled in 2004 in Shanghai, China, showing that electromagnetic suspension technology has in fact matured beyond expectations. The low-speed system using electromagnets, Linimo, also has been carrying 20,000 passengers per day in Nagoya, Japan, and has proven its advantages by offering a high level of reliability, considerable environmental appeal and lower maintenance costs. The Incheon International Airport Urban Maglev Demonstration Line in Korea and the Beijing and Changsa urban lines under construction in China will be opened in one or two years. On April 22, 2015, the Japanese superconducting magnet train L0 attained a running speed of 603 km/h, a record for any guided vehicle, and there are plans to operate this train over a route between Tokyo and Nagoya by 2027. These applications of magnetically levitated vehicles may prove that wheel-less transport could be a promising option as a new transportation mode in the future. On the other hand, as both the speed and the ride comfort performance of conventional wheel-on-rail vehicles have been considerably improved in recent years, the specific niche for magnetic vehicles in terms of speed as well as ride comfort is narrowing. It may be that magnetic vehicles are approaching a critical point that will determine their future viability. On the other hand, magnetic levitation technology is attracting strong interest for diverse applications in which the contact-free aspect is essential—examples include an extremely clean transfer system for LCDs and semiconductors, a rope-less elevator for skyscrapers and a hover board for entertainment purposes, and prototypes for these technologies have been demonstrated. With the rapidly increasing interest in its various forms and applications, there is an opportunity for engineers to play a dominant role in the development of contact-less or wheel-less operation systems.

The objective of this monograph is to discuss the principles of magnetic levitation and its operation in a way that can be understood by readers from various backgrounds. The authors also hope to promote a discussion that can lead to an enhancement of the current magnetic levitation system's competitiveness compared to conventional systems through innovation.

For ease of understanding and application, the three kinds of magnets, i.e., permanent, superconducting magnets, and electromagnet, in wide use are presented in a definitive and comprehensive manner with example cases and descriptions of the corresponding levitation concepts and configurations. The unique properties, advantages, and limitations, as well as significant problems of each magnetic levitation scheme, are discussed. In particular, railway applications are introduced chronologically and in more detail. The reader will find the book useful in imagining their own new concepts and identifying the basic design parameters. The majority of the content in this book can be understood by a reader who has studied university-level physics only, regardless of his or her major.

Much of this work, which has been supported by the MSIP, MOLIT, and KAIA, as well as the NST, was carried out in collaboration with the author's colleagues both at the KIMM (Korea Institute of Machinery and Materials) and from academic, research, and industry organizations.

Firstly, the authors would like to thank P.K. Sinha, the author of "Electromagnetic Suspension Dynamics & Control (1987)" for introducing them to the concept of magnetic levitation and providing the basis of this monograph by writing the first comprehensive and pioneering book in this area.

The authors would like to thank these colleagues for their contributions to the operational and experimental systems cited here. The contributions of Dr. Chang-Hyun Kim, Dr. Jae-Won Yim and Dr. Chang-Wan Ha and Dr. Han-Wook Cho to the magnetic system design, simulation, and test works given here are particularly acknowledged.

Technical details and figures quoted in the book are based on information available in the public domain, including via the Internet. In particular, the authors thank Dr. Byung-Chun Shin, a director of Center for Urban Maglev Program, for supplying information on Korea's ECOBEE, an urban magnetic levitation train. Professor Lin Guobin of Tongi University and Professor H. Osaki of the University of Tokyo are also gratefully acknowledged for supplying information on the status of magnetic trains in China and Japan.

Finally, the authors would also like to thank Dr. Peter-Juergen Gaede, Mr. Mizro Iwaya, and Dr. In-Kun Kim, who are developers, for their encouragement and guidance of this work thanks to their own extensive experiences with Transrapid, Linimo, and UTM developments, respectively.

Contents

Chapter 1
Introduction

To understand the context for the birth and evolution of maglev (short for magnetic levitation) technology, it is necessary to recall the economic situation and the road and air traffic situation the world faced in the 1960s. The global economy at the time was booming, a situation that led to congested road and air traffic. The cost of aviation was escalating due to rising fuel prices, and the conventional railway system was widely believed to have reached a peak. One of the reasons for the saturation of railways arose from the fact that rail joints limited the running speed of trains to below 200 km/h, as well as inducing shock, vibration and environmental noise when wheeled vehicles ran over the joints. It was not until the mid-1980s that high-speed trains that could travel in the range of 300 km/h appeared. To meet the growing demand for higher speed and greater comfort in intercity travel that was energy efficient and environmentally acceptable, a next-generation transportation system was needed—specifically, a new system capable of operating in the 300–500 km/h range. The reason is well illustrated by the equal travel time curve shown in Fig. 1.1. The curve suggests that, within the distance from 200 to 1500 km, high-speed trains that can travel in that speed range may be more attractive than other modes of transportation. Recently, with the operating speeds of wheel-on-rail systems approaching nearly 400 km/h, the speed gap between magnetic and wheel-on-rail systems has become smaller. However, some challenges still remain in operating wheel-on-rail systems at around 400 km/h. The maglev vehicle emerged as an alternative to the conventional railway system that could overcome both its speed limits and its environmental problems.

Suspension and propulsion are the basic functions required of any new generation vehicles. In magnetic levitation systems, magnetic fields are the media that can obtain the forces for the two functions without physical contact. The field is produced around a moving electric charge, e.g. electrons, and the interaction between the fields generates the mechanical forces of attraction or repulsion on the related objects. A coil carrying current and a permanent magnet are two of the very widely used methods to produce magnetic fields. The basic properties of a magnetic field are the generation of attraction and repulsion forces, depending on the same and opposing polarities facing each other, as shown in Fig. 1.2. These two kinds of forces generated by the magnetic fields are the primary forces involved in the suspension and propulsion of maglev systems.

© Springer Science+Business Media Dordrecht 2016
H.-S. Han and D.-S. Kim, *Magnetic Levitation*,
Springer Tracts on Transportation and Traffic 13,
DOI 10.1007/978-94-017-7524-3_1

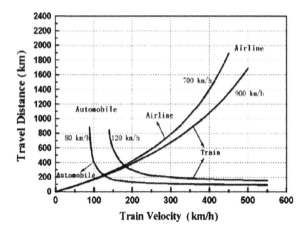

Fig. 1.1 Equal travel time curves [1]

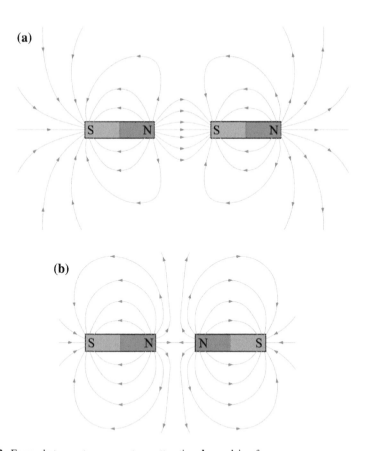

Fig. 1.2 Forces between two magnets: **a** attractive, **b** repulsive forces

The types of maglev can be broadly classified into electromagnetic attraction systems and electrodynamic repulsion systems. In the book, the levitation principles are simply categorized into attractive and repulsive systems to involve almost all the methods proposed. Though the terms levitation and suspension have a slightly different meaning, the former is used in this text to encompass both. Static permanent and superconducting magnets are representative repulsion systems. In addition to these, there are two methods for repulsive forces produced when the two magnets move over conductive sheets or coils, which are well-known EDS (Electrodynamic Suspension)s. The attractive type involves the use of electromagnets to maintain its separation between two objects by continuously adjusting the currents in the coil. In this method, a feedback control loop for levitation stabilization is required. Of course, there may be any number of combinations of repulsive and attractive methods above. Here, it is relevant to note that two kinds of linear motors are used for contact-less propulsion. LIM (Linear Induction Motor) is used mainly for low-speed propulsion, while LSM (Linear Synchronous Motor) can be used both for low and for high-speed propulsion.

The background of the need for maglev trains and its basic levitation methods are briefly described earlier. Using the magnetic cushion instead of wheel-on-rail, the contact-less vehicle has some advantages in its operational and environmental aspects. It has low vibration and noise, as well as least particles. More importantly, it is able to attain speeds that may not be possible or may be difficult in conventional rolling stock propelled by adhesion force between wheel and rail. While frictional force is limited by vehicle weight and friction coefficient, regardless of the power input into the traction motor, the linear motor for a maglev system, especially when mounted on the guideway rather than on board, can provide the vehicle with thrust sufficient to run at very high speeds. The upper speed is limited only by aerodynamic forces. For example, the superconducting maglev system L0 has a recorded speed of 603 km/h. Due to the speed and environmental advantages, the magnetic train could be expected to be a complementary mode of transportation to the conventional road and railway vehicles and aircrafts.

Nearly 40 years after work on maglev trains began, the first commercial maglev system Transrapid launched its operation in Shanghai, China in 2004. Key technical features are the use of electromagnets for suspension/guidance and LSM for high speeds up to 430 km/h over the Shanghai line (Fig. 1.3). The clearance of 10 mm between the magnets and guideway is maintained through electromagnets being controlled. While this system requires fine guideway tolerances due to a small air gap of 10 mm, this active suspension system can follow the guideway profile well, offering flexibility in terms of ride comfort and route layout. During its 10 years of problem-free operation, the system has proved that a maglev train using EMS can offer the high level of reliability and operational benefits inherent to maglev systems. This suggests that it will be a promising option for guided transportation in the future.

The Japanese maglev system L0 (Fig. 1.4) is a typical electrodynamic system using vehicle-borne superconducting magnets and ground coils installed in the side wall for levitation/guidance and propulsion. As the vehicle moves relative to the

Fig. 1.3 Transrapid in Shanghai, China [2]

Fig. 1.4 L0 series on Yamanashi maglev test line in Japan

coils, currents are induced in the ground coils, which in turn produce magnetic fields around them. The two magnetic fields produce levitation and lateral guidance forces. The main advantage of this system is that it requires no active levitation/guidance control. Recently, the vehicle hit a top speed of 603 km/h over the 42.8 km Yamanashi test track—a record for any guided vehicle. It is planned to be put into service over the route between Tokyo and Nagoya in 2027.

The Linimo (Fig. 1.5), an attraction type low-speed vehicle, has been carrying 20,000 passengers per day in Nagoya, Japan since 2005. Over 10 years of operation, its high level of reliability, considerable environmental appeal and low

Fig. 1.5 Linimo in Nagoya, Japan [3]

maintenance costs have proven its potential in urban and suburban applications. The Incheon International Airport Urban Maglev Demonstration Line in Korea (Fig. 1.6) and the Beijing and Changsa urban lines in China under construction will be opened in one or two years. These three vehicles are also attractive systems using electromagnets, and appear to present a promising option as a new transportation mode in ground transportation.

Owing to the technical success of maglev vehicles in railways, maglev is now attracting strong interest for diverse other applications where contact-free operation is a critical factor. Some of these are introduced below.

In the manufacturing processes in the display and semiconductor industries, ultra-cleanliness is one of the most important factors for economic production. Roller conveyors have been widely used for many years, and offer simplicity and cost effectiveness. However, this kind of contact-based system inevitably creates particles, which can lead to defects in the products. Speed is also another consideration. As an alternative to a roller conveyor, a maglev conveyor prototype shown

Fig. 1.6 ECOBEE in Incheon International Airport, Korea

in Fig. 1.7 was constructed using the same configuration as the urban maglev vehicles like ECOBEE and Linimo above. That is, a combination of electromagnets and LIM was employed. This conveyor can offer the ultra-clean transfer of LCDs or semiconductors at a relatively higher speed, resulting in improved productivity. There were also interesting attempts to make skateboards wheel-less through magnetic cushion (Fig. 1.8). They can hover over a conductive plate at a height of around 10 mm, allowing the rider to enjoy a smooth skateboarding experience. Based on electromagnet and LIMs, a PRT (Personal Rapid Transit) prototype, called skyTran, was produced (Fig. 1.9), which could be characterized as a guided taxi. This is a patented, high-speed, low-cost, elevated PRT system. It is claimed

Fig. 1.7 Maglev conveyor for LCDs [4]

Fig. 1.8 Magnetic skateboard

Fig. 1.9 PRT (Personal
Rapid Transit) using
electromagnets and LIMs [5]

that this concept is capable of carrying passengers in a fast, safe, green, and economical manner.

As mentioned above, maglev technology for vehicles has been developed to a fairly high level, and is attracting interest in various areas where contact-free operation is of critical importance. The purpose of the book is to give a comprehensive account of the design principles of magnetic levitation and its operations, referring to corresponding applications, including systems currently being researched. Permanent and superconducting magnets and electromagnets which are widely used are introduced in a definitive and comprehensive manner. The characteristic nature, advantages and limitations, as well as the outstanding problems of each magnetic levitation scheme are outlined. The contents of the book should be able to be understood by a reader who has an understanding of university-level physics, regardless of his or her major.

There are seven chapters in this monograph. Chapter 2 provides the fundamentals of electromagnetics, electronics, mechanics, control and measurements, and linear motors needed to enable the reader to easily understand the remaining chapters. Chapters 3 and 4 describe the levitation principles of permanent and superconducting magnetic systems and their main features. The topics covered here provide an introduction to system design principles rather than a detailed account of the physical properties of the magnets. Chapter 5, the largest section of the book, deals with the levitation principles of electromagnets. Unlike the two kinds of magnets described above, electromagnetic systems are inherently unstable and thus require feedback control loops. Because of their nature, a significant part is concerned with suspension stability. To implement electromagnetic systems, an analysis of system dynamics with control loops is needed considering appropriate electronics and measurements. In addition, the design and operational considerations of electromagnetic systems are presented, many of which are applicable both to permanent and superconducting systems. Chapters 3–5 are each followed by applications using each magnet. The configurations of each application may be useful for the reader to imagine their own maglev system. Chapter 6 is specifically dedicated to maglev trains only in service and under construction. The main features and status of each system are reviewed and discussed. A complete account of the systems is beyond the scope of this monograph because they are already well-known due to their long history of development. For detailed information about the magnetic trains cited in Chap. 6, it is recommended to refer to the relevant literature. Chapter 7 highlights the current research and development that is ongoing in various application areas, some of which is still in the prototype or concept stages. It is hoped by the authors that the reader will be able to see the potential of maglev in various applications, and will be inspired to imagine and realize their own maglev systems.

References

1. Yan Luguang (2009) The linear motor powered transportation development and application in China. Proc IEEE 97(11):1872–1880
2. Photo courtesy of ThyssenKrupp Transrapid GmbH/StoiberProductions. http://www.transrapid.de
3. Photo courtesy of Aichi Kosoku Kotsu. http://www.linimo.jp
4. Kim Chang-Hyun, Lee Jong-Min, Han Hyung-Suk, Lee Chang-Woo (2011) Development of a maglev LCD glass conveyor, Maglev 2011. Daejeon, Korea
5. Image courtesy of skyTran, Inc. http://www.skytran.us/images

Chapter 2
Fundamentals

2.1 Introduction

The design and implementation of maglev systems fundamentally involves the interdisciplinary concepts of electromagnetism, electronics, mechanical engineering, measurement and control. In consideration of this, the fundamentals of such disciplines are summarized to help the reader understand the remaining chapters with ease.

2.2 Electromagnetics

Magnetic levitation is realized through magnetic fields between magnetic objects. This section summarizes the source of the magnetic field and its nature as well as the related terminologies.

- **Magnetic field**: In a maglev system, the magnetic field is the medium that lifts vehicle systems weighing hundreds of tons and running at high speeds up to 600 km/h. What is the source of this magnetic field? It is moving electric charges. A moving charge q, at a velocity of $\vec{v}$, produces a magnetic field around it (Fig. 2.1). The magnetic field $\vec{B}$ is defined by

$$\vec{B} = \frac{\mu_0}{4\pi} \frac{q\vec{v} \times (\vec{r}/r)}{r^2} \qquad (2.1)$$

From Eq. (2.1), the magnitude of the field at any field $\vec{B}$ is given by

$$B = \frac{\mu_0}{4\pi} \frac{|q|v \sin\phi}{r^2} \qquad (2.2)$$

© Springer Science+Business Media Dordrecht 2016
H.-S. Han and D.-S. Kim, *Magnetic Levitation*,
Springer Tracts on Transportation and Traffic 13,
DOI 10.1007/978-94-017-7524-3_2

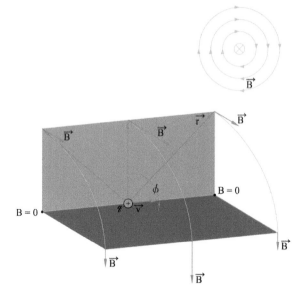

Equation (2.2) indicates that B is proportional to charge q and is inversely proportional to r^2. It is worth noting this relationship because magnetic forces are proportional to B. The constant μ_0 in the equation is described in other terms later.

Permanent magnets and solenoids also produce magnetic fields (Fig. 2.2). However, these magnetic fields are also produced by electric charges inside them made of magnetic materials. A magnetic field is formed by field lines. Field lines have no beginning or end, they always form closed loops. The field is spatial, as shown in Fig. 2.2. The field lines come from the North pole (N) and enter the South pole (S). The lines are continuous, and don't meet each other.

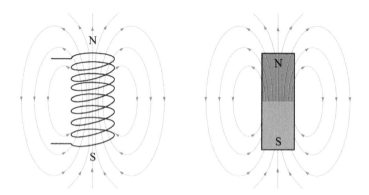

Fig. 2.2 Magnetic fields in solenoid (*left*) and permanent magnet (*right*)

- **Attraction and repulsive forces**: An attractive force is generated between two magnetized objects when the field lines go from one to another (Fig. 1.2a). Whereas if the same poles are facing each other, the field lines are pressured, resulting in a repulsive force between them (Fig. 1.2b). These two kinds of forces are the sources of magnetic levitation systems. For example, the Transrapid uses attractive forces.
- **Magnetic substance**: Some materials can be magnetized by an external magnetic field. The most common types of these materials are iron, nickel, cobalt and most of their alloys. Ferromagnetic materials have the strongest capacity for magnetism. Diamagnetic materials do not respond to an applied magnetic field. Iron is widely used as the material for magnetization in magnetic systems.
- **Change of magnetic field**: A magnetic field has a property to be formed in more permeable material. For example, the diamagnetic material glass does not influence the magnetic field, but if iron is put into the field, the field lines around it go through it, rather than air (Fig. 2.3). This is because the permeability of iron is higher than that of air. Using this property, magnetic field shielding and electromagnet, produced by winding iron with conductor carrying current, can be made.
- **Flux and flux density**: A series of field lines is called flux ϕ, and the number of the field lines is its value. A stronger magnetic field means a larger number of field lines, and consequently the value of flux becomes larger. That is, the flux value is the measure of a magnetic field's strength, and its unit is weber (Wb). 1 Wb indicates 10^8 field lines. Flux density B is defined as flux per unit area normal to magnetic field. Its unit is tesla (T). B is expressed as

$$B = \frac{\phi}{A} \left(\frac{Wb}{m^2} \right) \qquad (2.3)$$

Fig. 2.3 Change in the magnetic field with magnetic materials

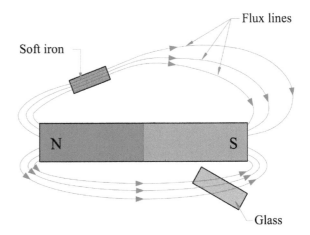

- **Electromagnetic**: If electrons with negative charges flow through a conductor, a magnetic field is produced around it (Fig. 2.4). This field is called an electromagnetic field, and the properties of the field are the same as the properties of a permanent magnet. The direction of the electromagnetic field is perpendicular to the wire, and moves in the direction the fingers of your right hand would curl if you wrapped them around the wire with your thumb in the direction of the current.
- **Permeability** (μ): Magnetic permeability represents the relative ease of establishing a magnetic field in a given material. The permeability of free space is called μ_0, and its value is $\mu_0 = 4\pi \times 10^{-7} H/m$. Relative permeability of any material $\mu_r = \mu/\mu_0$ compared to μ_0 is a convenient way to compare its magnetization. For steels, the relative permeabilities range from 2000 to 6000 or higher. Thus, if an iron core is wound by coils carrying currents, almost all of the flux produced by the coils goes through the iron core, not air, which has a smaller permeability than that of iron.
- **Reluctance** ($\mathscr{R}$): Reluctance is a magnetic resistance in materials, which is the counterpart of electrical resistance. It is defined as

$$\mathscr{R} = \frac{l}{\mu A} \tag{2.4}$$

where, l and A are the length and area of flux path.
- **Magnetomotive force**(m.m.f.): This is an analogy of voltage or electromotive force since it is the cause of magnetic flux in a magnetic circuit (Fig. 2.5). Magnetomotive force F_m is equal to the effective current flow applied to the core, that is

$$F_m = NI(ampere \cdot turns) \tag{2.5}$$

The relationship among electromagnetic properties is best illustrated by the magnetic circuit in Fig. 2.5. All of the magnetic field produced by the current will

Fig. 2.4 Magnetic field formed around a conductor carrying currents

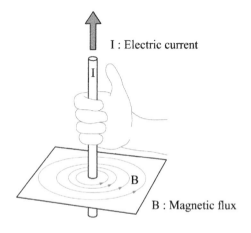

I : Electric current

B : Magnetic flux

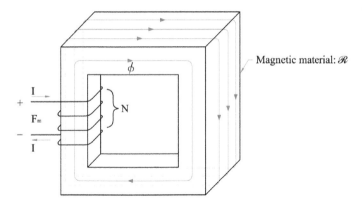

Fig. 2.5   A simple magnetic circuit

remain inside the core because the core's permeability is higher than air. And the flux of the magnetic circuit is defined as

$$\phi = \frac{F_m}{\mathscr{R}} \tag{2.6}$$

- **Electromagnet**: The electromagnet, which is most widely used in magnetic levitation systems, is a type of magnet in which the magnetic field is produced by an electric current. Electromagnets usually consist of a large number of closely spaced turns of wire that create the magnetic field. The wire turns are often wound around a magnetic core made from ferromagnetic materials. The magnetic core concentrates the magnetic flux and makes a more powerful magnet. The main advantage of an electromagnet over a permanent magnet is that the magnetic field can be quickly changed by controlling the amount of electric current in the winding. The direction of a magnetic field is dependent on the direction of the electric current. A common simplifying assumption satisfied by many electromagnets, which will be used in Chap. 5, is worth noting here, as it can reasonably be used in the design and analysis of electromagnets. The magnetic field of a U-shaped iron core electromagnet is shown in Fig. 2.6. The drawing shows a section through the core of the electromagnet except for the windings, which are shown in three dimensions for clarity. The iron core of the electromagnet (C) forms a closed loop for the magnetic flux, with two airgaps (G) in it. Most of the magnetic field (B) is confined to the core circuit. However, some of the magnetic field lines (B_L) take "short cuts" and do not pass through the entire core circuit, and thus do not contribute to the force exerted by the magnet; this is called "leakage flux." This also includes magnetic field lines that encircle the wire windings without entering the core. In the gaps (G), the magnetic field lines are no longer confined by the core, so they "bulge" out of the edges of the gap before bending back to enter the next piece of core material.

Fig. 2.6 Diagram of the
magnetic field of an
electromagnet

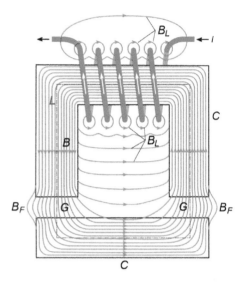

These bulges (B_F) are called "fringing fields" and reduce the strength of the magnetic field in the gaps. The blue line L is the average length of the flux path or magnetic circuit, and is used to calculate the magnetic field.

- **Magnetizing force**: The degree to which a magnetic field by a current can magnetize a material is called magnetizing force (H), and it is defined as m.m.f. per unit length of material. That is,

$$H = \frac{F_m}{l} = \frac{NI}{l} \quad (\text{Ampere} \cdot \text{Turn}/\text{meter}) \tag{2.7}$$

H is not related to a material's property. The magnetic flux (B) induced in material by H depends upon the nature of the material, and the relationship between H and B is defined by

$$B = \mu H \tag{2.8}$$

It is relevant here to relate all the parameters defined earlier as in Fig. 2.7.

- **Magnetic behavior of ferromagnetic materials**: Since the permeability of most ferromagnetic materials is not constant, the flux density B as a function of magnetizing field H is as shown in Fig. 2.8. Note that the flux in the materials is related linearly to the applied magnetomotive force in the unsaturated region, and approaches a constant value regardless of the magnetomotive force in the saturated region. Due to this behavior, the operational region of EMS systems should be located in the unsaturated region.
- **Hysteresis loop and residual flux**: Applying an alternating current to the windings on the core instead of a direct current with a frequency, the flux in the core traces out path *abcdeb* in Fig. 2.9. This is because the amount of flux

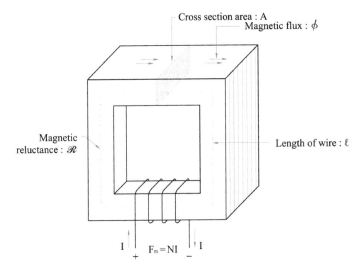

Fig. 2.7 Parameters determining H and B

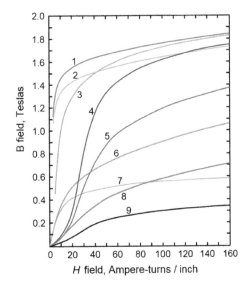

Fig. 2.8 Magnetization curves of 9 ferromagnetic substances; The substances are: *1* standard sheet steel, annealed, *2* silicon sheet steel, annealed, Si 2.5 %, *3* soft steel casting, *4* tungsten steel, *5* magnet steel, *6* cast iron, *7* nickel, 99 %, *8* cast cobalt, *9* Magnetite, Fe_2O_3

present in the core depends not only on the amount of current applied to the winding of the core, but also on the previous history of the flux in the core. This dependence on the preceding flux history and the resulting failure to trace flux paths is called hysteresis. Path *bcdeb* traced out in Fig. 2.9 as the applied current changes is called a hysteresis loop. If the frequency of an alternating current is changed, the path is also changed with a different residual flux. This property may lower the control performance at higher frequencies in a levitation system with electromagnets.

Fig. 2.9 The hysteresis loop traced out by the flux in a core when the alternating current is applied to it: **a** alternating current and **b** hysteresis loop

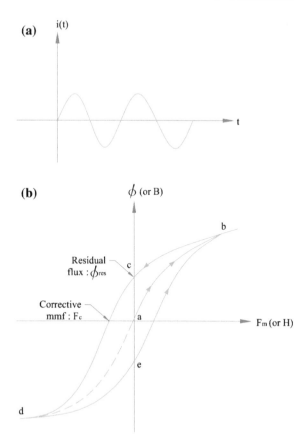

- **Electromagnetic induction**: When a conductor is exposed to a time varying magnetic field, a voltage is induced across it, as shown in Fig. 2.10. This can be mathematically described using Faraday's law of induction. Electric generators and motors as well as magnetic levitation systems are based on this electromagnetic induction. The polarity of induced voltage depends on the direction of relative motion. Expressed in the form of an equation for induced voltage e_{ind},

$$e_{ind} = -N\frac{d\phi}{dt}$$

(2.9)

The minus sign in Eq. (2.9) is an expression of Lenz's law to be described below.

- **Induced current**: If a conductor in Fig. 2.11 has an electrical resistance, an electrical current flows in the conductor. This current is called induced current i_{ind}.

- **Force on a current carrying conductor in a magnetic field**: If the directions of the flux lines from magnets and conductor are the same, the flux density

Fig. 2.10 Electromagnetic induction

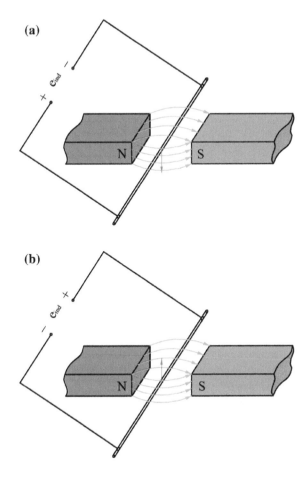

Fig. 2.11 Induced current

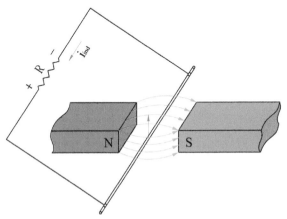

Fig. 2.12 Forces on current
carrying conductors in a
magnetic field

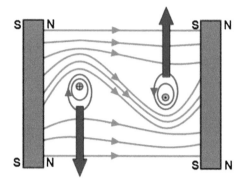

increases. In contrast, if the directions are opposite, the flux density decreases.
The resulting forces are exerted towards a weak magnetic pressure region from a
stronger pressure region, as shown in Fig. 2.12. This is the operating principle of
electric motors.

- **Lenz's law**: A time varying magnetic field induces voltage and corresponding
current in a conducting material described earlier. Lenz's law indicates that if an
induced current (eddy current) flows, its direction is always such that it will
oppose the change that produced it. Lenz's law is well applied to the permanent
magnet and moving conductive sheet in Fig. 2.13. The diagram of eddy currents
(*I*, red) induced in a conductive metal sheet (*C*) moving under a stationary
magnet (*N*) shows the directions of the induced currents. The magnetic field
lines (*B*, green) from the North pole of the magnet extend down through the
sheet. The increasing field at the leading edge of the magnet (left) causes the
currents to circle counterclockwise. Thus, based on Lenz's law they create their
own magnetic field directed upward which opposes the magnet's field, thus
exerting a drag and lift effect on the magnet. Similarly, at the trailing edge of the
magnet the decreasing magnetic field induces eddy currents that circle clock-
wise. This creates a magnetic field directed downward which attracts the

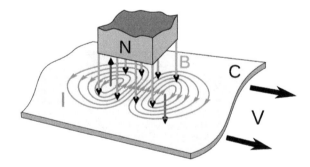

Fig. 2.13 Diagram of eddy currents (*I*, red) induced in conductive metal sheet (*C*) moving under a
stationary magnet

magnet, which also exerts a retarding force on the magnet. This is the principle of repulsive levitation in dynamic mode, with a moving magnet over a conductive sheet.

- **Inductance**: In electromagnetism and in electronics, inductance is the property of a conductor by which a change in current flowing through it induces (creates) a voltage (electromotive force) in both the conductor itself (self-inductance) and in any nearby conductors (mutual inductance). A changing electric current through a circuit that contains inductance induces a proportional voltage that opposes the change in current (self-inductance), as illustrated in Fig. 2.14. It is customary to use the symbol L for inductance. The unit for inductance is the henry (H). The relationship among the parameters for a coil with an Inductance L is defined as

$$V = IR - L\frac{dI}{dt} \qquad (2.10)$$

where

$$L = \frac{N^2 \mu A}{l}$$

Equation (2.10) indicates that Inductance opposes the applied voltage. This property to oppose building up currents influences the control performance of levitation systems with electromagnets.

- **Diamagnetic material**: Diamagnetic materials create an induced magnetic field in a direction that is opposite to an externally applied magnetic field, and are repelled by the applied magnetic field. Its magnetic permeability is less than μ_0. In most materials, diamagnetism is a weak effect, but a superconductor repels the magnetic field entirely, apart from a thin layer at the surface.

Fig. 2.14 Coil's reaction to increasing current

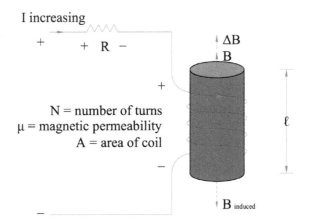

I increasing

+   + R −

+

N = number of turns
μ = magnetic permeability
A = area of coil

−

ΔB
B

ℓ

B induced

2.3   Electronics

In the implementation of maglev systems, some electrical equipment for power supply is combined with magnets and electric motors.

- **Chopper**: For the excitation of currents in an electromagnet, a chopper is used as a power amplifier. A chopper circuit is used to refer to numerous types of electronic switching devices and circuits used in power control and signal applications. A chopper is a switching device that directly converts fixed DC input to a variable DC output voltage. Essentially, a chopper is an electronic switch that is used to interrupt one signal under the control of another. For all the chopper configurations operating from a fixed DC input voltage, the average value of the output voltage is controlled by the periodic opening and closing of the switches used in the chopper circuit. The average output voltage can be controlled using different techniques. One of them is a PWM(pulse width modulation) technique. In the case of ECOBEE, the DC input voltage to a chopper operating in two quadrants is 350 V. A chopper circuit will be introduced in Sect. 5.5.10.
- **Inverter**: A power inverter, or inverter, is an electronic device or circuitry that changes direct current (DC) to alternating current (AC). The input voltage, output voltage and frequency, and overall power handling depend on the design of the specific device or circuitry. In maglev systems, VVVF (variable voltage variable frequency) inverters are used to drive linear motors and perform speed control.
- **Transformer and battery**: As in many electrical system, transformers and batteries are commonly used in maglev systems.

2.4   Mechanics

A maglev system consisting of mechanical and electrical components can be characterized as a dynamic system. In particular, the attractive maglev system with electromagnets may be treated as a forced vibration problem. The fundamentals of vibration analysis can be understood by studying the simple mass–spring–damper model. Indeed, even a complex structure such as a maglev train can be modeled as a "summation" of simple mass–spring–damper model. Thus, it may be useful to understand mechanical vibration for the design, analysis and control of a maglev system. Vibration is a mechanical phenomenon in which oscillations occur about an equilibrium point. There are two types of vibration, which are free and forced vibration. Free vibration occurs when a mechanical system is set off with an initial input and then allowed to vibrate freely. The mechanical system will then vibrate at one or more of its "natural frequencies" and damp down to zero. Forced vibration is when a time-varying disturbance (load, displacement or velocity) is applied to a mechanical system. For a maglev vehicle, the exciting electromagnet and uneven guideway surface profile are the main disturbances.

- **Free vibration without damping**: To start the investigation of the mass–spring–damper (Fig. 2.15), the damping is assumed to be negligible, and there is no external force applied to the mass (i.e. free vibration). The force applied to the mass by the spring is proportional to x (m) of deflection.

$$F_s = -kx \quad (N) \tag{2.11}$$

where k is the stiffness of the spring and has units of force/distance (N/m). The negative sign indicates that the force is always opposing the motion of the mass attached to it. According to Newton's second law of motion, the acceleration of the mass m (kg) is related to the force generated by the spring.

$$\sum F = F_s = ma = m\ddot{x} = m\frac{d^2x}{dt^2} \tag{2.12}$$

$$m\ddot{x} + kx = 0 \tag{2.13}$$

The solution of Eq. (2.13) has the form of $x(t) = A\cos(2\pi f_n t)$. This solution says that it will oscillate with a simple harmonic motion that has an amplitude of A and a frequency of f_n (Hz). f_n is called the undamped natural frequency. For the simple mass–spring system, f_n is defined as:

$$f_n = \frac{1}{2\pi}\sqrt{\frac{k}{m}} \tag{2.14}$$

In maglev systems, the natural frequency of the system may be determined by the relations described above. The concern is to obtain k from the magnet's force-airgap characteristic for excursions around its nominal position. This may be achieved through the calibration of experimental or analytical force-airgap data.

Fig. 2.15 Mass-spring system without damping

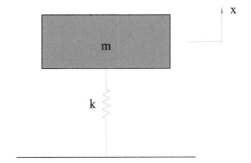

- **Free vibration with damping**: Force by any damping element with damping coefficient c (Ns/m) is determined by

$$F_d = -cv = -c\dot{x} = -c\frac{dx}{dt} \tag{2.15}$$

The resulting equation of motion of the system in Fig. 2.16 is expressed as:

$$m\ddot{x} + c\dot{x} + kx = 0 \tag{2.16}$$

This equation has the form of solution:

$$x(t) = Xe^{-\zeta\omega_n t}\cos(\sqrt{1 - \zeta^2}\omega_n t - \phi), \omega_n = 2\pi f_n \tag{2.17}$$

where
X	initial position
$\zeta = \frac{c}{c_c}$	damping ratio
$c_c = 2\sqrt{km}$	critical damping
ϕ	phase shift

The frequency in this case is called the "damped natural frequency", f_d and is related to the undamped natural frequency by the following formula:

$$f_d = f_n\sqrt{1 - \zeta^2} \tag{2.18}$$

What is often done in practice is to experimentally measure the free vibration after an impact (such as by a hammer) and then determine the natural frequency of the system by measuring the rate of oscillation as well as the damping ratio by measuring the rate of decay. The natural frequency and damping ratio are not only important in free vibration, but also characterize how a system will behave under forced vibration.

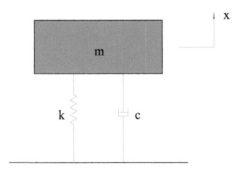

Fig. 2.16 Mass-spring-damper model

- **Forced vibration with damping**: The dynamic behavior of the mass-spring-damper system varies with applied force. If the external force is assumed to be

$$F = F_0 \sin(2\pi f t) \tag{2.19}$$

Summing all the forces on the mass, the equation of motion leads to:

$$m\ddot{x} + c\dot{x} + kx = F_0 \sin(2\pi f t) \tag{2.20}$$

The solution of the equation in steady-state can be written as:

$$x(t) = X \sin(2\pi f t) \tag{2.21}$$

The solution indicates that the mass will oscillate at the same frequency, f, of the applied force, but with a phase shift ϕ. The amplitude of the vibration "X" is defined by the following formula.

$$X = \frac{F_0}{k} \frac{1}{\sqrt{(1-r^2)^2 + (2\zeta r)^2}} \tag{2.22}$$

where "$r = f/f_n$" is defined as the ratio of the harmonic force frequency over the undamped natural frequency of the mass–spring–damper model. The phase shift, ϕ, is defined by $\phi = \arctan(2\zeta r/(1-r^2))$. The plot of these functions in Fig. 2.17, called "the frequency response of the system," presents one of the most important features in forced vibration. In a slightly damped system, when the forcing frequency nears the natural frequency ($r \approx 1$), the amplitude of the vibration can become extremely high. This phenomenon is called resonance (subsequently, the natural frequency of a system will often be referred to as the resonant frequency). In any maglev system, a time-varying disturbance with a speed that excites a resonant frequency is referred to as a critical speed. If resonance occurs in a mechanical system it can be very harmful. Consequently,

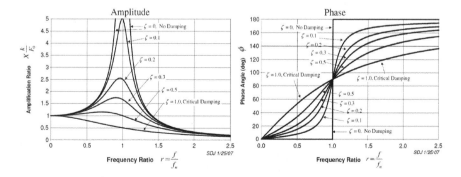

Fig. 2.17 Frequency response of a mass-spring-damper system

one of the major reasons for vibration analysis is to predict when this type of resonance may occur, and then to determine what steps to take to prevent it from occurring. As the amplitude plot shows, adding damping can significantly reduce the magnitude of the vibration. The magnitude can also be reduced if the natural frequency can be shifted away from the forcing frequency by changing the stiffness or mass of the system. If the system cannot be changed, perhaps the forcing frequency can be shifted. These vibration characteristics and stabilization techniques can be equally applied to maglev systems.

Multiple degrees of freedom systems (MDOF) and mode shapes: In more complex systems, the system must be discretized into more masses which are allowed to move in more than one direction—adding degrees of freedom. Equations of motions of a MDOF can be in a matrix form, as in the following:

$$[\mathbf{M}]\{\ddot{x}\} + [\mathbf{C}]\{\dot{x}\} + [\mathbf{K}]\{x\} = \{f\} \tag{2.23}$$

where $[M], [C],$ and $[K]$ are symmetric matrices referred to respectively as the mass, damping, and stiffness matrices. The matrices are $N \times N$ square matrices where N is the number of degrees of freedom of the system. If there is no damping and applied force (i.e. free vibration), the solutions of the system may be assumed to have the form of:

$$\{x\} = \{X\}e^{i\omega t} \tag{2.24}$$

With this solution, the system equations of motion Eq. (2.24) becomes

$$\left[-\omega^2[\mathbf{M}] + [\mathbf{K}]\right]\{X\}e^{i\omega t} = 0 \tag{2.25}$$

Since $e^{i\omega t}$ cannot be zero, the equation reduces to the following.

$$\left[[\mathbf{K}] - \omega^2[\mathbf{M}]\right]\{X\} = 0 \tag{2.26}$$

This equation becomes an eigenvalue problem. It can be put in the standard format by pre-multiplying the equation by $[M]^{-1}$.

$$\left[[\mathbf{M}]^{-1}[\mathbf{K}] - \omega^2[\mathbf{M}]^{-1}[\mathbf{M}]\right]\{X\} = 0 \tag{2.27}$$

If $[M]^{-1}[K] = \{A\}$ and $\lambda = \omega^2$, the resulting eigenvalue problem is obtained as:

$$[[\mathbf{A}] - \lambda[\mathbf{I}]\{X\}] = 0 \tag{2.28}$$

The solution to the problem results in N eigenvalues (i.e. $\omega_1^2, \ldots, \omega_N^2$), where N corresponds to the number of degrees of freedom. The eigenvalues provide the natural frequencies of the system. When these eigenvalues are substituted back into the original set of equations, the values of $\{X\}$ that correspond to each

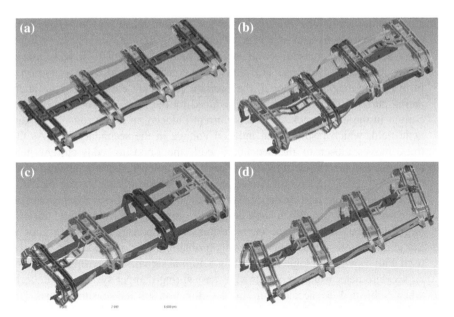

Fig. 2.18 Vibration mode shapes

eigenvalue are called the eigenvectors. These eigenvectors represent the mode shapes of the system. Since manually deriving the eigenvalues and eigenvectors for a large scale system may be too time consuming or involved, FEM programs are widely used. As an example, Fig. 2.18 shows a set of vibration modes of the experimental maglev vehicle, in Sect. 5.3, obtained by a FEM program. Understanding the vibrational characteristics of MDOFs is more important for the stabilization of electromagnetic systems. This topic will be discussed in Chap. 5.

2.5 Control and Measurement

Especially in electromagnetic suspension systems, taking the measurements of position, velocity and acceleration for the feedback control loop is the first stage from an operational viewpoint. Once the needed signals are measured, after appropriate signal processing, they are input to a control loop. The fundamentals in levitation controller design and its implementation are briefly outlined in this section.

- **Sensors**: For maglev systems, contactless transducers are required for the measurement of position, velocity, and acceleration. In practice, position sensors and accelerometers are mostly used, especially in vehicles. The criteria for selecting a suitable transducer include bandwidth, robustness and stability under all operating conditions, linearity over the operating range, and immunity from noise, radiation and stray magnetic fields. Among the various contact position sensors, inductive transducers are outlined here because of their wide use in low-speed maglev vehicles. This transducer consists of two coils wound on a form of non-magnetic material, the primary coil being energized by an alternating current. The magnetic field produced by the primary coil induces eddy current in the metal track. The eddy current induced, in turn, produces a magnetic field, which results in an induced voltage in the secondary coil. As the track moves closer to the secondary coil, the opposing eddy-current field increases, and the output of the secondary coil is reduced. Through the use of an appropriate feedback control circuit, the output of the transducer can be made linear, from 0 to 20 mm clearance. The bandwidth of this device is dependent on the carrier frequency. When such devices are used with ferromagnetic track as target material, however, unacceptable transient responses for steps or gaps at rail joints may result. To overcome this problem, a pair of gap sensors may be used with a switching logic for selecting a normal signal. The sensing element of an accelerometer usually consists of a mass-spring-damper system that deflects when subjected to acceleration in the direction of its sensitive axis. The deflection of the seismic mass is a linear function of applied acceleration, according to Newton's second law, within the constraints imposed by the natural frequency and the damping ratio of the seismic system. The natural frequency and damping ratio of the devices are determined by mass, stiffness of spring equivalent, and damping in the device. Piezoelectric accelerometers are also available for the feedback control loop. In these accelerometers, the sensing element is a small disc of piezoelectric material that generates an electric charge when it is compressed or extended by a mass. Primary selection criteria for accelerometers are frequency response and range. For most accelerometers, bandwidth is dependent on the range of operating acceleration levels. In maglev vehicles, the range of 3 to 5 g may be chosen, though acceleration during normal operation is not expected to be higher than 0.1 g. Consequently, in selecting adequate sensors one must consider the factors of the particular application.

- **Transfer function:** The transfer function of a linear system is defined as the ratio of the Laplace transformation of the output variable to the Laplace transform of the input variable, with all initial conditions assumed to be zero. The transfer function of system $G(s)$, shown in Fig. 2.19, represents the relationship describing the dynamics of the system under consideration. It is expressed as follows:

$$G(s) = \frac{Laplace\ transform\ of\ output\ variable\ y(t)}{Laplace\ transform\ of\ input\ variable\ r(t)} = \frac{Y(s)}{R(s)}$$

$$G(s) = \frac{b_0 s^m + b_1 s^{m-1} + \cdots + b_{m-1} s + b_m}{a_0 s^n + a_1 s^{n-1} + \cdots + a_{n-1} s + a_n} \tag{2.29}$$

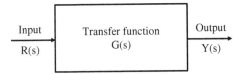

Fig. 2.19 Transfer function

- **Pole and zero**: The poles of the transfer function defined above are roots of the
  denominator polynomial, and the zeros are roots of the numerator polynomial.
  Poles significantly influence both steady-state and transient response. The effects
  of poles can be summarized as follows:

  - A system is stable if a pole is located in the left half-plane, whereas it is
    unstable if the pole is in the right half-plane.
  - When a pole lies in the left half-plane, the response decays rapidly as it
    moves away from the imaginary axis.
  - When a pole lies in the right half-plane, the response diverges rapidly as the
    pole moves away from the imaginary axis.
  - The frequency of a response increases as a pole moves away from the real
    axis.

  The influence of zero on the response of the system is transient. The influence
  can be summarized as follows:

  - Zero does not significantly affect the response when zero is placed far from
    the imaginary axis.
  - Overshoot increases as zero moves closer to the imaginary axis while zero is
    in the left hand-plane.
  - When overshoot on the down side appears, the magnitude increases as zero
    moves closer to the imaginary axis.

- **Bode plot**: A Bode plot is a very useful way to study a frequency response of
  the transfer function with analytic or experimental results. The transfer function
  in the frequency domain is

$$G(s = j\omega) = |G(\omega)|e^{j\phi(\omega)} \qquad (2.30)$$

  where the units are decibels (dB). The logarithmic gain in dB and the angle
  $\phi(\omega)$ can be plotted versus the frequency ω. Control bandwidth, Gain margin
  and Phase margin related to the Bode plot are important parameters for the
  stabilization of electromagnetic suspension. As an example, Bode plot for
  Butterworth filter is given in Fig. 2.20.

- **Filters**: The feedback control system of electromagnetic systems needs signal
  processing to accurately derive airgap, velocity or acceleration from the sensor
  output signals. In signal processing, a filter is a device or process that removes
  some unwanted component or feature from a signal. Filtering is a class of signal

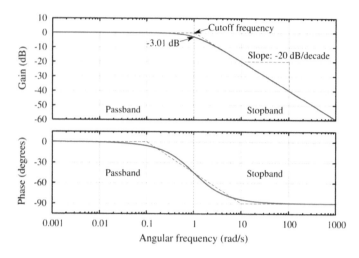

Fig. 2.20 Exemplary Bode plot for Butterworth filter

processing, the defining feature of filters being the complete or partial sup-
pression of some aspect of the signal. Most often, this means removing some
frequencies and not others in order to suppress interfering signals and reduce
background noise. The frequency response of filters can be classified into a
number of different bandforms, as shown in Fig. 2.21, describing which frequency bands the filter passes (the passband) and which it rejects (the stopband).
Cutoff frequency is the frequency beyond which the filter will not pass signals. It
is usually measured at a specific attenuation, such as 3 dB. Roll-off is the rate at
which attenuation increases beyond the cut-off frequency. Transition band is the
(usually narrow) band of frequencies between a passband and stopband. Ripple
is the variation of the filter's insertion loss in the passband. The parameters such
as cutoff frequency must be chosen based on considerations for the particular
system. Deriving accurate signals is the primary work in levitation stabilization.

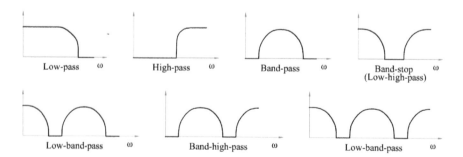

Fig. 2.21 Filter bandform diagrams

- **Numerical differentiation**: In the implementation of a control loop, one may need derivatives of a signal, like airgap, using numerical analysis techniques. Taylor series expansions are most commonly employed to derive these. Since those are well known formulas, the resulting ones are just given here. The well-known linearization technique is the same as the following centered finite-divided difference formulas of first derivatives. Because errors in approximations depend on step size (time step), the step size must be verified by evaluating errors. The following numerical differentiation formula are well illustrated in Fig. 2.22.

Fig. 2.22 Graphical depiction: **a** forward, **b** backward, and **c** centered finite-divided-difference approximations of the first derivative

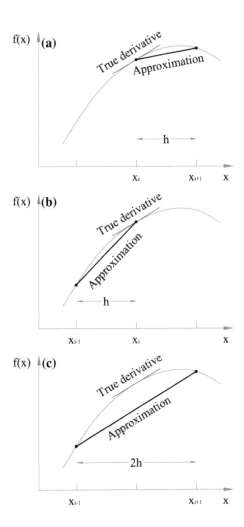

- Forward finite-divided difference formulas:
First derivative

$$f'(x_i) = \frac{f(x_{i+1}) - f(x_i)}{h}$$

Second derivative

$$f''(x_i) = \frac{f(x_{i+2}) - 2f(x_{i+1}) + f(x_i)}{h^2}$$

Third derivative

$$f'''(x_i) = \frac{f(x_{i+3}) - 3f(x_{i+2}) + 3f(x_{i+1}) - f(x_i)}{h^3}$$

- Backward finite-divided-difference formulas:
First derivative

$$f'(x_i) = \frac{f(x_i) - f(x_{i-1})}{h}$$

Second derivative

$$f''(x_i) = \frac{f(x_i) - 2f(x_{i-1}) + f(x_{i-2})}{h^2}$$

Third derivative

$$f'''(x_i) = \frac{f(x_i) - 3f(x_{i-1}) + 3f(x_{i-2}) - f(x_{i-3})}{h^3}$$

- Centered finite-divided difference formulas:
First derivative

$$f'(x_i) = \frac{f(x_{i+1}) - f(x_{i-1})}{2h}$$

Second derivative

$$f''(x_i) = \frac{f(x_{i+1}) - 2f(x_i) + f(x_{i-1})}{h^2}$$

Third derivative

$$f'''(x_i) = \frac{f(x_{i+2}) - 2f(x_{i+1}) + 2f(x_{i-1}) - f(x_{i-2})}{2h^3}$$

- **Integration of differential equations**: Dynamics equations describing maglev systems are a type of differential equation, i.e. $y'(x,t) = f(x,t)$, and thus the solutions for them are usually derived through a numerical integration scheme. Though there is a wide numerical integration scheme, only Euler's method, a one-step method, and the Runge-Kutta method, a multistep method, are summarized because they are most frequently used in maglev systems. Because errors in approximations depend on step size (time step h), the step size must also be verified by evaluating errors.

  – Euler method:

$$y_{i+1} = y_i + f(x_i, y_i)h, f(x_i, y_i) = dy/dx \tag{2.31}$$

  – Fourth-order Runge-Kutta method:

$$y_{i+1} = y_i + \frac{1}{6}(k_1 + 2k_2 + 2k_3 + k_4)h \tag{2.32}$$

  where

$$k_1 = f(x_i, y_i)$$
$$k_2 = f(x_i + \frac{1}{2}h, y_i + \frac{1}{2}k_1 h)$$
$$k_3 = f(x_i + \frac{1}{2}h, y_i + \frac{1}{2}k_2 h)$$
$$k_4 = f(x_i + h, y_i + k_3 h)$$

2.6 Linear Motors

Once a system is levitated, linear motors are naturally chosen to propel it. Of the various available types, LIM (linear induction motor)and LSM(linear synchronous motor) appear to be the most suitable for maglev systems.

Linear motors can be conceptually described as unrolled versions of the familiar rotary machine. This means that the design and operation features of linear motors are the same as those of rotary motors. For this reason, it is recommended to refer to the literature on rotary machines. The application of linear motors to vehicles is given in subsequent chapters. Three possible configurations of LIMs in Fig. 2.23 can be used in any maglev system. Stator (primary) or rotor (secondary) can be installed on a moving object or fixed guideway, and vice versa. These configurations offer considerable simplicity in determining the guideway installation cost. The principle of a LIM in use for urban maglev vehicles is conceptually

Fig. 2.23 Transformation of
rotary induction motors into
linear versions: **a** drag cup
rotary, **b** single-sided linear:
short stator, **c** single-sided
linear: long stator,
d double-sided linear with
two short stator windings

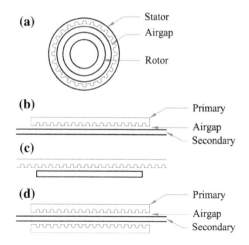

demonstrated in Fig. 2.24. The interaction between the traveling magnetic field of
the 3-phase winding on-board and the field produced by eddy-currents in the aluminum plate on guideway, induced by the traveling magnetic field, gives thrust to
accelerate vehicle. The magnitude of the thrust can be controlled with varying the
current and excitation frequency in the 3-phase winding (primary). The reactive
power of LIM is usually high, and goes up as the airgap increases, reducing its
power factor. This makes these motors well suited for attraction type system with
airgaps around 15–20 mm. LSM consists of multiphase iron- or air-cored winding
and field magnets, which are permanents or electromagnets. For the combination of
iron-cored winding and an electromagnet as a field magnet, the principle of thrust is
illustrated in Fig. 2.25. The three-phase winding can propel a vehicle that travels in
synchronism with the electromagnetic wave. The slip speed is zero, but there is a
current angle (θ, Fig. 2.25), which indicates the position of the vehicle with respect
to the travelling wave, the thrust being maximum when $\theta = 90°$. By controlling the
current and frequency of the primary excitation, as a function of the current angle,

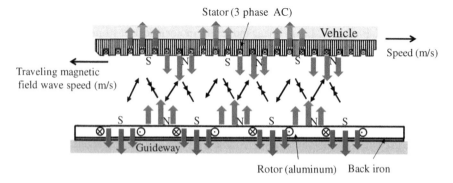

Fig. 2.24 Principle of a LIM for use in urban maglev vehicle

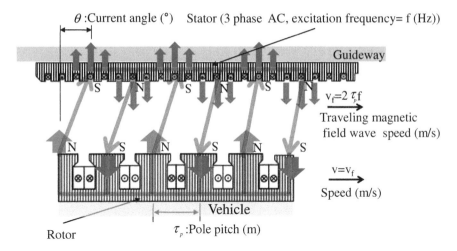

Fig. 2.25   Principle of a LSM for use in high-speed maglev vehicle

the desired longitudinal speed/acceleration may be obtained. In addition, since θ also influences the vertical force, it may be used to introduce additional damping into the suspension loops. The winding configuration of the primary (armature) and the secondary (field coil) as well as wavelength/pole pitch are key important parameters of this system which is estimated to have an overall efficiency of around 85 %. Although ideally suited for large airgaps, the LSM may also be used with the attraction system in which the airgaps are likely to be below 25 mm. In maglev vehicles currently in service, LIM is being used for low-to-medium speeds, with LSM being used for high speeds.

Chapter 3
Permanent Magnet

3.1 Introduction

A permanent magnet creates its own persistent magnetic field once magnetized by any external magnetic field. The main feature of a permanent magnet is that a continuous supply of power is not needed, and thus a maglev system that employs a permanent magnet has a very simple configuration and lower maintenance costs. On the other hand, it is vulnerable to loss of field due to heating, and the oscillatory motion may be slowly diminished due to its lower damping property. Allowing for the development of a permanent magnet's increasing magnetic strength, the use of permanent magnets for levitation may be expected to attract more interest in various applications. In addition, since most people are familiar with the use of permanent magnets in everyday life, levitation that uses them is easy to imagine. In this chapter, the material properties of permanent magnets are first introduced. Then, the magnetic levitation principles based on them, both in static and dynamic modes, are given, along with corresponding applications. A novel configuration of permanent magnets for increasing flux density, the Halbach array, is outlined. The Halbach array is being widely used both in vehicles and in industry. However, it is beyond the scope of this chapter to cover the detailed properties and manufacturing process of permanent magnets.

3.2 Material

The suitable permanent magnet materials are well identified by the magnetic hysteresis loop in Fig. 3.1. When an external magnetic field (magnetomotive force, H) is applied to a ferromagnetic material, the material becomes magnetized (magnetic

© Springer Science+Business Media Dordrecht 2016
H.-S. Han and D.-S. Kim, *Magnetic Levitation*,
Springer Tracts on Transportation and Traffic 13,
DOI 10.1007/978-94-017-7524-3_3

Fig. 3.1 Magnetic hysteresis loop

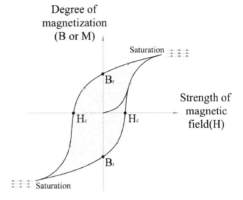

Table 3.1 Primary magnetic properties of four materials

Material	B_r (tesla)	H_c (tesla)	Specific density
Nd–Fe	1.25	1.11	7.4
Sm_2CO_{17}	1.12	0.69	6.4
Ferrite	0.44	0.29	5.0
Alinco	1.15	0.16	7.3

flux *B or M*) and will stay magnetized infinitely. The relationship between field strength *H* and magnetization *B* is not linear in such materials. As *H* increases, *B* increases following the initial magnetization curve, approaching an asymptote called magnetic saturation. If *H* is now reduced to 0, after following a different curve, the magnetization is offset from the origin by an amount called the residual flux B_r. If the *H-B* relationship is plotted for all strengths of an applied magnetic field the result is a hysteresis loop called the main loop. From this loop, it can be identified that the materials having larger B_r are suitable for permanent magnets. In addition to B_r, reducing self-demagnetization leads to a larger H_c being required to make *B* 0. Consequently, the larger B_r and H_c materials are appropriate for permanent magnets. Four commonly available magnetic materials, as examples, are given in Table 3.1 with associated key properties B_r and H_c.

3.3 Static Repulsive and Attractive Modes

Levitation using permanent magnets in static mode is based simply on static repulsive or attractive forces between two magnets. A static repulsive levitation system is shown in Fig. 3.2a. The system's configuration is very simple, while some wheels are needed to restrict its movements in vertical or lateral direction. Attractive type systems are illustrated in Fig. 3.2b. Conceptually, both of these two

Fig. 3.2 Possible modes of
levitation by using permanent
magnets in static mode:
a repulsive and b attractive
system

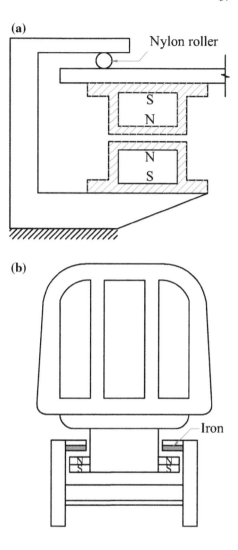

systems are very simple. Based on this attractive system, a commercial system, M-Bahn (Fig. 3.2b), was developed and put into service in July 1991 in Berlin, although the system was removed due to the reunification of Berlin. Only 85 % of the M-Bahn vehicle weight is supported by magnetic levitation, with the balance being supported by traditional wheels. For propulsion, it used a long stator linear motor. Currently, however, there are no static permanent magnet vehicles in service or under construction. Although this system may need additional gears for movement restrictions and secondary suspension for ride comfort in vehicles, with the development of higher strength magnets, they may have potential for use in applications like conveyor systems for lightweight cargo.

3.4 Halbach Array

Permanent magnets have some limitations on magnetic flux density and difficulty in assembly due to attracting ferromagnetic materials such as steel. To overcome these limitations, a novel arrangement of permanent magnets, the Halbach array, was proposed by Klaus Halbach to increase magnetic field strength. An example of a Halbach array is shown in Fig. 3.3, in which the arrows indicate the orientation of each piece's magnetic field. This array would give a strong field underneath, while the field above would cancel. The reason for this flux distribution can be intuitively visualized using Mallinson's original diagram, as shown in Fig. 3.4. The diagram

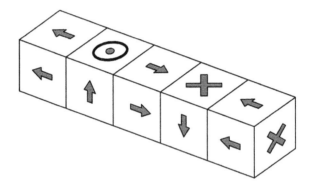

Fig. 3.3 Simple model of a Halbach array showing the orientation of the fields of magnetic components

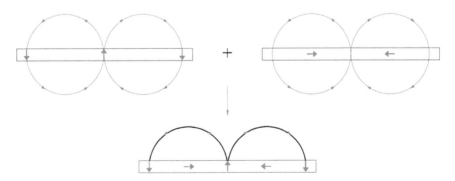

Fig. 3.4 Mallinson Halbach array figure

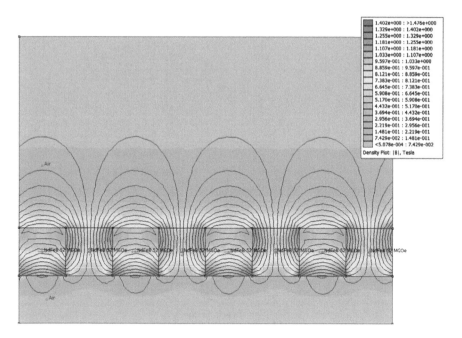

Fig. 3.5 Field lines around a Halbach array

shows the field from a strip of ferromagnetic material with alternating magnetization in the y direction (top left) and in the x direction (top right). Note that the field above the plane is in the same direction for both structures, but the field below the plane is in opposite directions. The effect of superimposing both of these structures is shown in the figure. The crucial point is that the flux will cancel below the plane and reinforce itself above the plane. This superimposing effect is clearly visualized in Fig. 3.5, which indicates that the flux on one side is reinforced. Because of this feature, most maglev vehicles using permanent magnets are based on a Halbach array. A maglev system using a Halbach array in static mode has been proposed [1, 2]. The forces to lift the vehicle at a gap of 3–8 cm are produced by the repulsive forces between the on-board Halbach array and reaction magnets. While the configurations of these attractive and repulsive systems with permanent magnet in static mode are very simple, they may be unstable in any direction and need reaction magnets on the track, requiring an additional mechanism to restrict their movements as well as protection to avoid attracting ferromagnetic objects around them (Fig. 3.6c).

Fig. 3.6 Repulsive maglev
system with Halbach arrays:
a flux distribution,
b configuration, and
(**c**) prototype [1, 2]

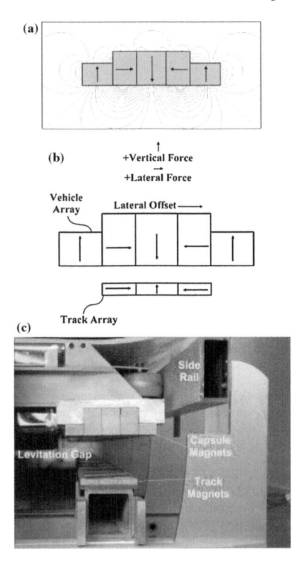

(a)

(b)   +Vertical Force
                 +Lateral Force

Vehicle
Array   Lateral Offset ⟶

Track Array
(c)

3.5 Dynamic Repulsion

One of the main limitations of a static levitation system with permanent magnets is
the difficulty in their operation due to the persistency of their magnetic fields.
Repulsive forces can be generated when a permanent magnet moves over a conducting sheet (Fig. 3.7), such as an aluminum sheet. The repulsive force magnitude
depends on the relative velocity between a moving magnet and the conducting sheet.
This kind of levitation that is dependent on speed is called electrodynamic suspension (EDS). If a permanent magnet moves relative to a conducting sheet, its magnetic

Fig. 3.7 Moving permanent magnet: **a** dynamic interaction and **b** levitation force

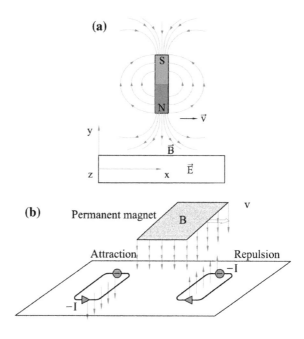

field induces currents in the sheet, and the currents in turn produce magnetic fields. The two magnetic fields produce repulsive forces by the interaction between them. To illustrate this levitation principle, it is assumed that a permanent magnet moves at $\vec{v}$ relative to a conducting sheet fixed to ground, as shown in Fig. 3.7a. Magnetic flux density of the magnet $\vec{B}$ in air moves relative to the conducting plate. According to Lenz's law, a magnetic field opposing $\vec{B}$ should be created by the induced currents in the plate. That is, the electromotive force required to create the magnetic field is induced in the sheet according to Faraday's law. The induced electric field $\vec{E}$ (V m^{-1}) is derived using Faraday's law as

$$\vec{E} = \vec{v} \times \vec{B} \qquad (3.1)$$

By this electric field $\vec{E}$, the current $\vec{J}$ (current density, A m^{-2}) flows in the conducting material with electrical conductivity σ.

$$\vec{J} = \sigma\vec{E} \qquad (3.2)$$

This induced current is called eddy current. Substituting Eq. (3.1) into Eq. (3.2) and assuming the magnet is moving only along the x axis, the eddy current induced is derived as follows:

$$\vec{J} = \sigma \times (\vec{v} \times \vec{B}) = \sigma \times \begin{bmatrix} \vec{i} & \vec{j} & \vec{k} \\ v & 0 & 0 \\ B_x & B_y & B_z \end{bmatrix} = J_x\vec{i} + J_y\vec{j} + J_k\vec{k} \qquad (3.3)$$

$$J_x = 0, J_y = -\sigma v \cdot B_z, J_z = \sigma v \cdot B_y \qquad (3.4)$$

From Eqs. (3.3) and (3.4), it is indicated that the magnitude of eddy current is proportional to σ, v and flux densities B_y and B_z.

Once eddy current is induced in the conductive sheet, Lorentz force is generated, and it is expressed as Eq. (3.5).

$$\vec{F} = \vec{J} \times \vec{B} = \begin{bmatrix} \vec{i} & \vec{j} & \vec{k} \\ 0 & J_y & J_z \\ B_x & B_y & B_z \end{bmatrix} = F_x\vec{i} + F_y\vec{j} + F_z\vec{k} \qquad (3.5)$$

where $F_x = (J_y \cdot B_z - J_z \cdot B_y)$, $F_y = J_z \cdot B_x$, $F_z = -J_y \cdot B_x$. F_x corresponds to magnetic drag while F_y corresponds to levitation force. In addition, note that F_z presents also. The force F_x represents the thrust required to propel the permanent magnet. Because these force equations for a moving magnet are the source of levitation in electrodynamic levitation systems, a more rigorous derivation of them in terms of flux density, resistivity and velocity is given in Eq. (3.6).

$$\begin{aligned} F_x &= -\sigma v (B_z \cdot B_z + B_y \cdot B_y) \\ F_y &= \sigma v (B_x \cdot B_y) \\ F_z &= \sigma v (B_x \cdot B_z) \end{aligned} \qquad (3.6)$$

Equation (3.6) suggests that increasing B_x and decreasing B_y are preferred to reduce drag force, while F_y exists in as large a force as possible. The levitation principle of this type is conceptually shown in Fig. 3.7b, in which the repulsive force is greater than the attractive force, resulting in levitation. The velocity dependency of the levitation and drag forces are well demonstrated in Fig. 3.8, which shows that levitation force increases with velocity, approaching a saturation value with $v \to \infty$. The drag force, on the other hand, has been observed to be inversely proportional to the velocity, after being increased with speed. These characteristics may be desirable to allow repulsive levitation by using moving permanent magnets.

One of the ways to increase the strength of magnetic flux density B_x is to use a multi magnet arrangement, with the polarities of the magnets being arranged alternatingly and closely (Fig. 3.9). Owing to closeness of two magnets and their polarities, a flux linkage is formed between two magnets, which has the effect of strengthening B_x. This flux strengthening effect depends on the distance between the two magnets and the directions of their polarities.

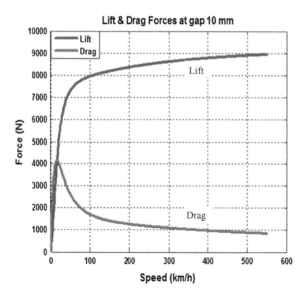

Fig. 3.8 Example of levitation/drag force-speed characteristics of a moving magnet over conductive plate

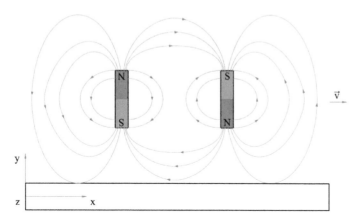

Fig. 3.9 Flux linkage between two permanent magnets

The Halbach array described earlier may be such an optimized magnet arrangement that it strengthens flux density in airgap. The levitation mechanism with Halbach array instead of permanent magnets is the same as that of Fig. 3.7. The primary advantage of using the Halbach array is a considerable improvement in the lift to weight ratio (Fig. 3.10).

A roller rig was constructed to study the speed dependency of the levitation and drag forces with Halbach array (Fig. 3.11). The Halbach array consists of 9 magnets

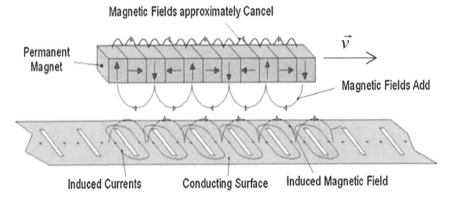

Fig. 3.10 Electrodynamic levitation with a moving Halbach array magnet [3]

Fig. 3.11 A roller rig to measure the levitation and drag forces of the Halbach array with increasing speed

in two rows. The permanent magnet material is Nd–Fe–B with $Br = 1.1$ T and $\mu_r = 1.0446$, and its dimensions are $50 \times 50 \times 50$ mm. The thickness of the conductive plate of aluminum is 35 mm. The separation between magnets and conductive plate is 40 mm. The measured lift and drag forces are given in Fig. 3.12, which shows the typical pattern in ordinary permanent magnets. At 100 km/h and above, the lift force increases smoothly with increasing speeds, approaching a saturated value. On the other hand, the drag force decreases from around 15 km/h. As mentioned earlier, this drag force is the required thrust. Here, a comparison of the levitation force results of the experiment to those of the numerical simulation shows a significant difference, the latter being almost 60 % greater than the former. This considerable difference suggests that 3D FEM analysis may be used to predict the forces created by a Halbach array (Fig. 3.13).

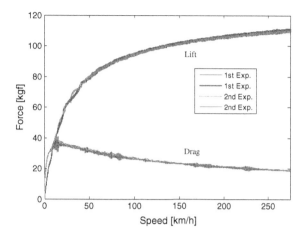

Fig. 3.12 Lift and drag forces-speed characteristics measured

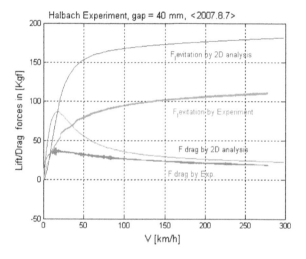

Fig. 3.13 Comparison of experimental levitation and drag forces to those from 2D FEM analysis

These repulsive levitation systems using moving permanent magnets are inherently stable, but may require relatively greater thrust to overcome magnetic drag forces. Many high-speed magnetic trains that employ a Halbach array have been conceptually proposed, but none are yet in service. For low-speed trains, GA has built the first full-scale working maglev system in the U.S., which operates on a 400-ft. test track at GA's main campus in San Diego [4]. The levitation technology is referred to as "inductrack" and uses high (1.4 T) NdFeB permanent magnet cubes ($5 \times 5 \times 5\,cm^2$) arranged in a double Halbach array configuration, as seen in Fig. 3.14a. One of the advantages of this configuration is that the magnetic field is "focused" on the track and tends to cancel on the passenger side. The array has been

(a)

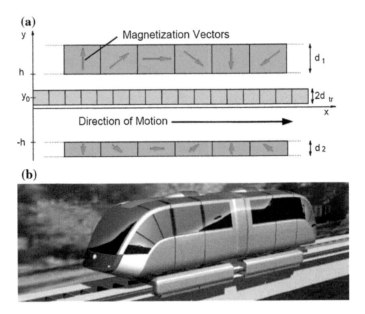

(b)

Fig. 3.14 GA's levitation system using **a** double Halbach array and **b** its vehicle [4]

configured to provide a nominal air gap of 25 mm, providing the potential for less stringent guideway tolerance requirements. Since this is an electrodynamic levitation system, the vehicles initially ride on wheels and then start levitating at a speed of 3–4 m/s, depending on the weight.

Magplane, which is shown in Fig. 3.15, is another proposed Halbach array maglev train for high speeds [5]. Halbach arrays of neodymium-iron-boron material are mounted underneath the vehicle. It is claimed that 5900 kg magnets are capable of lifting a weight of 32,000 kg.

One of the limitations of repulsive levitation that involves moving permanent magnets above is the need for translational relative movements. There is a way to obtain relative velocity by rotating magnets rather than translational movements. This technique is called magnetic wheel. The nature of levitation and drag forces are the same as those from translational relative movements; the only difference is that a rotational movement is a replacement for a translational movement. A simple configuration of a magnetic wheel for levitation is given in Fig. 3.16, which shows a wheel with magnets alternatingly arranged that is rotated by a electric motor [6–8]. With this configuration, an object can be levitated regardless of relative translational movement. While these systems are relatively energy inefficient, if the magnet arrangement is optimized, they might be viable in particular applications. Some applications will be introduced in Chap. 7. The key design parameters of this system are the geometry and the number of magnets, the rotational speed, and the

Fig. 3.15 Magplane, a
maglev vehicle for high
speeds, using Halbach
array [5]

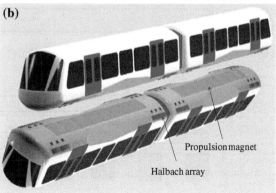

Fig. 3.16 Configurations of
magnetic wheels [6]

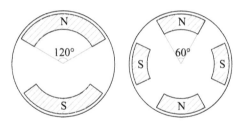

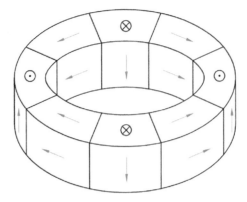

Fig. 3.17 Halbach array for magnetic wheels [9, 10]

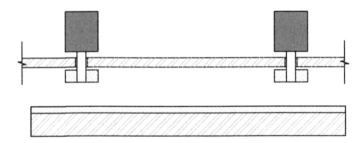

Fig. 3.18 The magnetic wheels on-board the mover [9, 10]

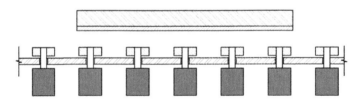

Fig. 3.19 The magnetic wheels on the floor [9, 10]

radius of the wheel. The use of a Halbach array for magnetic wheels was patented, as seen in Figs. 3.17, 3.18 and 3.19 [9, 10].

This magnetic wheel is placed on the on-board mover or floor. When magnetic wheels are installed on the floor, the rapid temperature rise in the conductive plate, if a mover does not move, may be a serious problem due to the eddy currents in it.

References

1. Long G (2008) Design of a small-scale prototype for a stabilized permanent magnet levitated vehicle. Maglev 2008, San Diego, USA
2. Fiske OJ (2006) The Magtube low cost maglev transportation system. Maglev 2006, Dresden, Germany
3. Source: http://sunlase.com/
4. Gurol S, Baldi R, Bever R (2004) Status of the General Atomics low speed urban maglev technology development program. Maglev 2004, Shanghai, China
5. Montgomery DB (2004) Overview of the 2004 magplane design. Maglev 2004, Shanghai, China
6. Jung KS, Shim KB (2010) Noncontact conveyance of conductive plate using omni-directional magnet wheel. Mechatronics 20:496–502
7. Park JH, Baek YS (2008) Design and analysis of a maglev planar transportation vehicle. IEEE Trans Magn 44(7):1830–1836
8. Fujii N, Hayashi G, Sakamoto Y (2000) Characteristics of magnetic lift, propulsion and guidance by using magnet wheels with rotating permanent magnets. In: Industry applications conference 2000 of IEEE, vol 1, pp 257–262
9. Patent: KR 1012156300000 (2012) Magnetic levitation system having Halbach array
10. Patent: KR 1011740920000 (2012) Magnetic levitation system having Halbach array

Chapter 4
Superconducting Magnet

4.1 Introduction

Levitation using superconducting magnets has come a long way since Arkadiev's experiment in the early 1940s, resulting in the superconducting maglev train L0 series for the route between Tokyo and Osaka at 505 km/h. Recently, several prototypes were built to transfer semiconductors using superconducting magnets in a clean environment. With the advances in high-temperature superconducting materials with more manageable proportions, this type of levitation can be expected to be worthy of further consideration. The three methods for levitation based on diamagnetism, flux pinning effect and inductive repulsion and attraction of super-conducting systems are outlined, along with their associated applications.

4.2 Superconductivity

Some conductors lose their electrical resistance completely below a particular temperature, becoming ideal current-carrying conductors. This phenomenon of superconductivity is influenced by temperature and the ambient magnetic field. This behavior is illustrated in Fig. 4.1, where the particular temperature is referred to as the critical temperature (T_c). Usually, the critical temperature (T_c) is below $-240\,°C$, e.g. $-268.8\,°C(4.2\,K)$ for metallic mercury. With the discovery of crystal structures of high-temperature ceramic superconductors with up to $T_c = 138\,K$, the levitation scheme through superconducting magnets is attracting interest in the areas of transportation and automation. Once an electric current is set up in a supercon-ducting loop, the current flows and induces a magnetic field persistently as long as the temperature is kept below critical temperature, making the superconducting coil behave like a permanent magnet.

© Springer Science+Business Media Dordrecht 2016
H.-S. Han and D.-S. Kim, *Magnetic Levitation*,
Springer Tracts on Transportation and Traffic 13,
DOI 10.1007/978-94-017-7524-3_4

Fig. 4.1 Critical field and
resistivity of superconductors

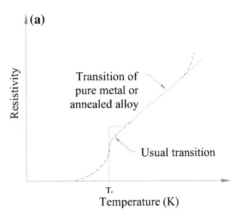

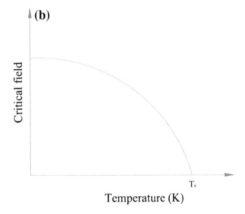

The ambient magnetic field is another key parameter for superconductivity.
When a conductor is subjected to an external magnetic field above a particular
strength, the superconductivity is destroyed. This is called the critical field (H_c). H_c
depends on the temperature, and becomes 0 at T_c. The relationship between T_c and
H_c is given by

$$H_c = H_0 \left[1 - \frac{T^2}{T_c^2}\right] \tag{4.1}$$

where, H_0 is the critical magnetic field at 0 K, H_c is the critical magnetic field at T K
and T_c is the critical temperature at zero magnetic field. The transition between
normal conduction and superconduction of materials is reversible with the changing
temperature or the ambient field of strength. Materials with a high critical magnetic
field and current density are appropriate superconductors. Superconducting materials are divided into two groups:

- Type-I (soft): superconductors with very low critical magnetic fields. The familiar aluminum with H_c around 17×10^{-3} T is a soft superconductor.
- Type-II (hard): superconductors with relatively high critical fields. The alloy niobium-germanium-aluminum is a hard conductor with H_c around 10 T.

The behaviors of the above two types are shown in Fig. 4.2. In Type-I materials, there is only one critical field at below the critical temperature, at which the transition between two states occurs. In superconducting state at below the Hc, perfect diamagnetism is exhibited. That is, the superconductors may be considered perfect diamagnets, since they expel all fields (except in a thin surface layer) due to the Meissner effect. This diamagnetism is the key phenomenon in the context of the magnetic field. However, its use is technically difficult due to its instability. In contrast, Type-II has the upper and critical fields depending on temperatures. As seen in Fig. 4.2b, an interesting and useful state exists between lower and upper critical fields. This state is a mixed state (or vortex state), which has properties both of superconducting (diamagnetism) and normal conducting. Some magnetic fields are repelled, and some penetrated it. The amount of penetration of fields increases with the increasing strength of fields up to the upper critical field. Most compounds and high-temperature superconductors are Type-II, which is most widely used. The rejection of the magnetic field by superconductors forms the basis of superconducting maglev systems.

4.3 Diamagnetism

Diamagnetism may be used for levitation. If a superconducting material (Type-I) is subjected to an external magnetic field at below the critical field, the external field is perfectly rejected by the material, resulting in levitation. Figure 4.3 shows the levitation of a superconducting object through diamagnetism. This levitating method is conceptually very simple. However, there may be the need for a balancing configuration for stabilization. Currently, there is no maglev train in service or under development based on this principle.

4.4 Flux Pinning

One of the main limitations of the above approach is its unstable nature. In contrast, the flux pinning effect of Type-II superconductors can provide stable levitation (Fig. 4.4). Flux pinning is the phenomenon in which a superconductor is pinned in space above a magnet. The superconductor must be a Type-II superconductor due to the fact that Type-I superconductors cannot be penetrated by magnetic fields. The act of magnetic penetration is what makes flux pinning possible. At higher temperatures, the superconductor allows magnetic flux to enter in quantized packets

Fig. 4.2 Behavior of Type-I
and Type-II superconductors
in a magnetic field: **a** Type-I
and **b** Type-II

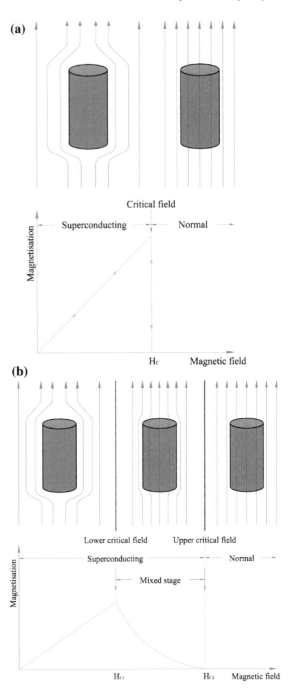

Fig. 4.3 Levitation of
superconducting object by
diamagnetism (Meissner
effect) [1]

surrounded by a superconducting current vortex. These sites of penetration are
known as flux tubes. The number of flux tubes per unit area is proportional to the
magnetic field, with a constant of proportionality equal to the magnetic flux
quantum. At low temperatures, the flux tubes are pinned in place and cannot move.
This pinning is what holds the superconductor in place, thereby allowing it to
levitate. This phenomenon is closely related to the Meissner effect, though with one
crucial difference—the Meissner effect shields the superconductor from all magnetic fields causing repulsion, unlike the pinned state of the superconductor disk,
which pins flux and the superconductor in place. Consequently, the superconductor
can be levitated by diamagnetism for lifting force and flux pinning for stabilization.

A maglev vehicle based on flux pinning called SupraTrans has been built
(Fig. 4.5). The system configuration of SupraTrans for levitation consists of the

Fig. 4.4 Illustration of flux
pinning and the magnetic
fields that are associated with
it

Fig. 4.5 The maglev vehicle, named SupraTrans, using the flux pinning effect [2]

permanent magnetic part installed on the guideway and a high-temperature superconductor block (HTSL) within cold vessels (cryostats) at −196 °C (Fig. 4.6).
During refrigeration to below critical temperature, the superconductor memorizes
the penetrated magnetic flux from the permanent magnets. The pinned flux holds
the superconductor (HTSL) above the permanent magnets, providing lateral guidance. For levitation, the rejection of the fields induced by the permanent magnets
generates lifting forces on the superconductor. Cryostat is refrigerated with liquid
nitrogen. The technical features of the SupraTrans are listed in Table 4.1.

Fig. 4.6 Superconductor
with pinning centers within
the magnetic field of a
3-pole magnetic rail
(cross-section) [2]

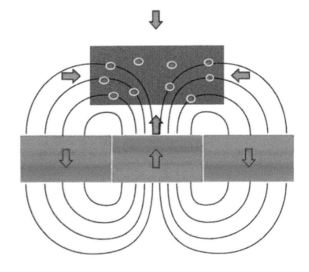

Table 4.1 Specifications of the SupraTrans [2]

Item	Value
Guideway	
Track gauge	1000 mm
Track length	80 m
Radius of curve	6.5 m
Mean flux density in the air space	0.6 T
Separation between conductor and track	13 mm
Mechanical airgap below cryostat	10 mm
Vehicle	
Length	2500 mm
Width	1200 mm
Total mass (including 2 persons)	600 kg
Superconductor material	YBCO
Max. power for thrust	3.4 kW
Max. thrust force	600 N
Max. acceleration	1 m/s^2
Max. velocity	20 km/h

A full scale maglev vehicle using the flux pinning effect named the Maglev-Cobra has been constructed and is undergoing test runs (Figs. 4.7 and 4.8). Repulsive forces are generated due to the flux pinning effect between the super-conducting material within the vehicle-borne cryostat and Halbach arrays installed on the guideway [3, 4]. Each cryostat contains 24 YBCO bulks of $32 \times 64 \times 12$ mm in 2 rows. The bulks are placed within the copper block, which is isolated through a vacuum chamber. The copper block is filled with liquid nitrogen (LN_2) with cavity. Halbach arrays consisting of NdFeB permanent magnets are installed on the guideway. The main advantage of this system is that it is passively stable, providing both levitation and guidance.

Fig. 4.7 Maglev-Cobra using flux pinning effect [5]

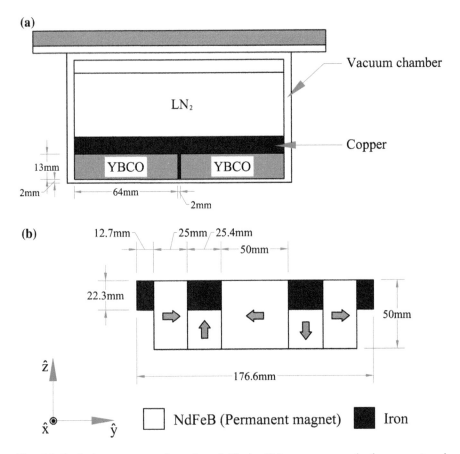

Fig. 4.8 Levitation system configuration of Maglev-Cobra: **a** superconducting magnet and **b** Halbach array guideway [3]

4.5 Electrodynamic Levitation

Permanent magnet conductive levitation systems with a moving magnet and conductive plate have been introduced in Chap. 3. Their operation principles can be equally applied to superconducting systems for levitation in dynamic mode. Therefore, excepting for the strength of the magnetic field produced by the superconductor, most of its features may be the same as those of a permanent magnet system. This levitation principle is best illustrated by the roller rig shown in Figs. 4.9 and 4.10. The superconducting coil faces the aluminum sheet wound around a rotating wheel, and is within the cryostat filled with liquid nitrogen. If the roller rotates, the relative tangential speed of the coil with respect to the aluminum sheet varies, as in translational movements. Due to the relative velocity, the very high strength of the magnetic field induces a circulating eddy current in the

Fig. 4.9 A superconducting coil within cryostat

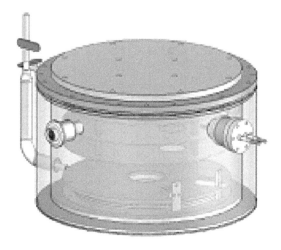

Fig. 4.10 Roller rig for experimentation of levitation and drag forces

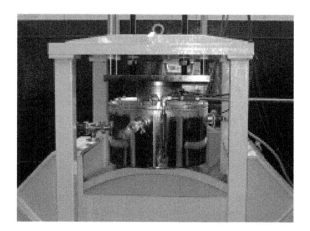

conductive sheet. This eddy current in turn will induce a magnetic field, which, as per Lenz's law, will oppose the magnetic field created by the moving superconducting magnet, producing repulsive and drag forces on the superconducting magnet. The levitation and drag forces measured are given in Fig. 4.11. For convenience, the drag force is scaled by 3×. The behaviors of two forces are the same as those of a permanent magnet. The levitation force increases with increasing speeds, approaching a saturation point. On the other hand, the drag force decreases from at about 20 km/h.

The levitation force-speed characteristics indicate that at lower speeds, the force may be not sufficient for lifting the magnet and its associated gear off the conducting sheet. The speed when net lift force is positive is called critical speed. In this type of electrodynamic levitation system, an auxiliary wheeled suspension is

Fig. 4.11 Levitation and drag
force at different air gaps (*FL*:
levitation force, *Fd*: drag
force, Current I = 20 A, ■: air
gap = 30 mm, O: air
gap = 35 mm, ∇: air
gap = 40 mm)

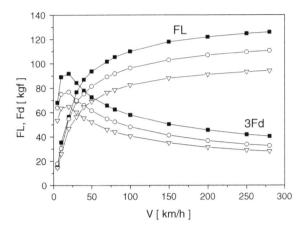

needed for operations at below critical speeds. This critical speed is dependent on
vehicle weight and the magnetic field strength in the airgap. Once the system is
levitated, the levitation will be sustained as long as the critical field and temperature
requirements are satisfied.

This system requires a considerable airgap flux density ($\geq 1T$) to be viable, and
consequently Type-II superconducting materials are chosen, with aluminum used
for the guideway. With such a combination, the stronger magnetic fields induced by
the conducting sheet are rejected by the superconductor's fields, yielding a
self-stabilizing levitation force in steady-state. Only D.C. superconducting magnets
are suitable for the above system, because hard superconductors with A.C. excitation have relatively high hysteresis loss. In practice, the amplitude of persistent
current in the superconducting loop will be reduced because there is a very small
residual resistance in it. In addition, the coolant around superconducting loops
should be replaced to keep the temperature of coils below critical temperature.
Furthermore, a certain amount of ohmic loss in the aluminum plate generates heat
and a magnetic drag force on the moving magnet. The nature of decreasing drag
force with increasing speed makes the electrodynamic levitation methods more
viable, particularly in maglev trains running at higher speeds. The lift to drag force
may be used as a qualitative measure of the viability of this maglev system. This
drag force and its associated properties with dynamics will be discussed later. For a
more easy understanding of the electrodynamic levitation of a moving superconducting coil above the aluminum plate, the repulsion force may be expressed in
terms of the magnetic interaction between the moving coil and an imaginary coil of
opposite polarity located at an equal distance below the plate (track) (Fig. 4.12) [6].
Using this assumption, the image force $F_l(t)$ may be expressed as

$$F_l(t) = [i(t)]^2 \frac{dM(x, \hat{z}, t)}{d\hat{z}} \qquad (4.2)$$

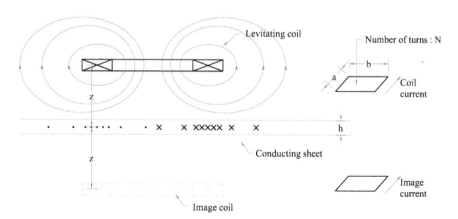

Fig. 4.12 Magnetic field and induced current distribution of a D.C. superconducting coil moving over an aluminum conducting plate

where $i(t)$ is the current in the conducting coil and $M(x, \hat{z}, t)$ is the mutual inductance between the true and image coils. The distance between the coils is represented by $\hat{z} = 2z$, and $x(t)$ is the longitudinal position of the coil from a reference point. The time derivative of $x(t)$ gives the longitudinal velocity v.

The magnetic field interaction between two coils creates two forces. One is the levitation force $F_l(t)$ and drag force (longitudinal) due to ohmic losses in the aluminum conducting sheet. The behaviors of the two forces with respect to increasing velocity are already presented earlier and in the preceding chapter. $F_l(t)$ increases with the longitudinal velocity, approaching $F_l(t)$ with $v \to \infty$. On the other hand, the drag force is inversely proportional to the velocity, and the exact relationship is related by

$$
\begin{aligned}
F_d(t) &\propto \frac{1}{v} \quad \text{for} \quad h < \delta \\
&\propto \frac{1}{v} \quad \text{for} \quad h > \delta
\end{aligned}
\tag{4.3}
$$

where h is the thickness of the conducting plate and δ is the skin depth $\delta = \lambda / \pi \mu_0 \sigma v$. The resulting ratio of the lift and drag forces $[F_l(t)/F_d(t)]$ improves with increasing velocity. The general expressions for lift and drag forces on a current-carrying conductor moving relative to a conducting object are known as [6]

$$
F_{ls}(v, z, t) = F_{ls} \left[\frac{v^2}{v^2 + w^2} \right] = F_{ls} \left[1 - \frac{1}{\left(1 + \frac{v^2}{w^2}\right)} \right]
\tag{4.4}
$$

$$F_{ds}(v, z, t) = F_{ls}\left[\frac{vw}{v^2 + w^2}\right] = F_{ls}(v, z, t) \tag{4.5}$$

where F_{ls} is the image force on the single conductor, and may be expressed as (with Eq. 4.2)

$$F_{ls} = \frac{\mu_0 I^2}{4\pi z}$$

I being the constant coil current and z the steady-state height of the coil above the conducting sheet. The constant w is defined by $w = 2/\mu_0\sigma h$. The force equations for a single conductor suggest that the ratio of lift/drag forces may be increased by choosing a conducting sheet thickness lower than the skin depth of the conducting sheet material. This conclusion on the ratio of lift/drag forces applies to magnets consisting of rectangular coils. Analytical derivation of the two forces is difficult due to the time history dependency of the two forces. It has been observed by researchers that the coil geometry does not have any influence on the drag force given by Eq. (4.5), while a closer estimate of the lift force is given by the empirical relationship [6]. As an alternative to analytical derivations of the two forces, the use of a commercial electromagnetic field analysis program may be a practical way to obtain the two forces, both in transient and steady-state modes. A 2D or 3D FEM may be used to simulate the two forces in a way that has good agreement with experimental results. The equations of motion of the moving coil, a magnet, above the conducting plate may be written as

$$\text{vertical motion: } m\ddot{z}(t) = -mg + F_l[v(t), z(t)] \tag{4.6}$$

$$\text{longitudinal motion: } m\ddot{x}(t) = F_p - F_d[v(t), z(t)] - f_{ad} \tag{4.7}$$

where m is the mass of the magnet(coil assembly), F_p is the propulsion force and f_{ad} is the aerodynamic drag forces on the magnet. A commonly used expression for aerodynamic force on a running vehicle is

$$f_{ad} = (1/2)C_d LA \,\rho v^2 = k_{ad}v^2$$

where L and A are length and cross-sectional area of the vehicle, ρ is the air density, C_d is the drag coefficient desired to be chosen by experimentation. Thus for a steady-state (v_0, z_0), i.e. $\ddot{z}(t) = \ddot{x}(t) = 0$, the equations of motion are rewritten as

$$F_{l0}(v_0, z_0) = mg \tag{4.8}$$

$$F_{d0}(v_0, z_0) = F_p - k_{ad}v_0^2 \tag{4.9}$$

Linearized equations of a moving magnet may be written as

$$m\Delta \ddot{z}(t) = \left.\frac{\partial F_l(v,z,t)}{\partial v}\right|_{(v_0,z_0)} \Delta \dot{x}(t) + \left.\frac{\partial F_l(v,z,t)}{\partial z}\right|_{(v_0,z_0)} \Delta z(t)$$

$$= \alpha_{lv}\Delta \dot{x}(t) + \alpha_{lz}\Delta z(t) \tag{4.10}$$

$$m\Delta \ddot{x}(t) = \left.\frac{\partial F_d(v,z,t)}{\partial v}\right|_{(v_0,z_0)} \Delta \dot{x}(t) + \left.\frac{\partial F_d(v,z,t)}{\partial z}\right|_{(v_0,z_0)} \Delta z(t)$$

$$= \alpha_{dv}\Delta \dot{x}(t) + \alpha_{dz}\Delta z(t) \tag{4.11}$$

Taking Laplace transformation of Eqs. (4.10) and (4.11), the characteristic equation of the electrodynamic levitation system may be expressed as

$$\left(s^2 - \frac{\alpha_{lz}}{m}\right)\left(s + \frac{\alpha_{dv}}{m}\right) + \frac{\alpha_{lv}\alpha_{dz}}{m^2} = 0 \tag{4.12}$$

where the constant slopes at the equilibrium point (v_0, z_0) are given by [6]

$$\alpha_{lz} = -\frac{mg}{z_0}, \alpha_{dz} = -\frac{wmg}{z_0}$$

$$\alpha_{lv} = -\frac{2mg}{v_0}\frac{1}{\left(1 + \frac{v_0^2}{w^2}\right)}$$

for n = 1 : single conductor

$$\alpha_{dv} = \frac{wmg}{v_0^2}\left[\frac{\left(1 - \frac{v_0^2}{w^2}\right)}{\left(1 + \frac{v_0^2}{w^2}\right)}\right] + 2k_{ad}v_0 \text{ for single conductor}$$

$$\tag{4.13}$$

With these constants, Eq. (4.12) may be rewritten as

$$\left[s^2 + \frac{g}{z_0}\right]\left[s + \frac{wg}{v_0^2}\left\{\frac{1 - \frac{v_0^2}{w^2}}{\left(1 + \frac{v_0^2}{w^2}\right)}\right\} + \frac{2k_{ad}v_0}{m}\right] = 0 \tag{4.14}$$

The first term is related to vertical motion and second term horizontal (longitudinal) motion. This implies that the vertical oscillatory motion of the superconducting levitation system is almost uncoupled from its horizontal motion. From a levitation viewpoint, the vertical dynamic characteristic is more important and thus needs to be more discussed. Equation (4.14) suggests that the vertical dynamics of a superconducting levitation system are influenced only by the height z. As an example, for an airgap of 10 cm, Eq. (4.14) gives the frequency of undamped oscillation about 1.6 Hz as follows.

$$\omega_0 = \frac{g}{z_0} \cong 10 \text{ rad/sec} \rightarrow f_0 = 1.6 \text{ Hz} \qquad (4.15)$$

Although an exact nonlinear analysis of the system described by Eqs. (4.6) and (4.7) would predict a damped vertical response, the amount of damping inherent in the electrodynamic system is not significant. For ride comfort, the vertical oscillations of around the frequency could not be suppressed by conventional suspension systems in vehicles. This is because the secondary suspension between the passenger compartment and the bogie in railway vehicles is adjusted to have a vertical motion at around 1 Hz, and to reduce its amplitude. For this reason, some form of damping needs to be incorporated into the electrodynamic levitation system. This requirement of some amount of damping is equally also applied to permanent magnet levitation systems in dynamic mode. This feature is one of the limitations of the electrodynamic levitation principle. In contrast, the controlled electromagnetic levitation systems to be introduced in Chap. 5 can achieve considerable damping through levitation control loops. There are two ways to generate some damping between the magnets and the conducting sheet. These are briefly discussed in the following sections. Configurations of full-scale vehicles using superconducting levitation systems in dynamic mode are given in Fig. 4.13, and these were constructed in the 1970s. The superconducting levitation systems are used both for vertical lifting and lateral guidance with auxiliary wheels. The superconducting coils are mounted underneath the vehicle and the continuous conductive metal sheet is fixed to the ground. Currently, there are no maglev vehicles in service or being operated that use these configurations. One of the reasons may be an unacceptable drag force, lowering the ratio of the lift to drag forces.

One of the methods of improving the ratio of the lift to drag forces is to use rectangular levitation coils for lifting and rectangular coils for the reaction surface (continuous conductive sheet). One of the objectives of this configuration is to increase the efficiency in magnetic fields contributing to levitation forces by concentrating them. This concept is shown in Fig. 4.14, and was applied in the Japanese ML-100 and ML-500 in the 1970s. If the vehicle with rectangular superconducting coils moves above the rectangular ground coils, the magnetic fields are induced in the ground coils, and in turn the induced fields oppose the fields of the superconducting coils, yielding levitation and drag forces. The basic levitation principle of this configuration is the same as that introduced earlier.

A modification of the configuration in Fig. 4.14 is given in Fig. 4.15, in which the two coils are arranged in such a manner that they remain perpendicular. For this configuration, the flux linkage, induced voltage and induced current are related as in Fig. 4.15. MLU-001 in Fig. 4.16 was built and operated based on the above concept (Fig. 4.16).

The latest and most advanced version of using superconducting magnets in dynamic mode is the null-flux system employed by L0 vehicle (Fig. 4.17). This scheme was proposed by Powell and Danby in the 1960s, who suggested that

Fig. 4.13 Possible combinations of superconducting levitation systems

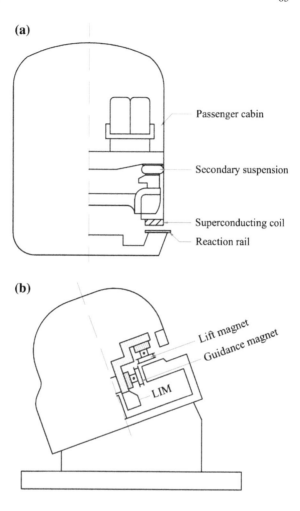

(a)

Passenger cabin

Secondary suspension

Superconducting coil

Reaction rail

(b)

Lift magnet

Guidance magnet

LIM

superconducting magnets could be used to generate the high magnetic pressure needed. Null flux systems work by having coils on the side wall that are exposed to a magnetic field from an on-board superconducting coil, but are wound in 8-shaped and similar configurations such that when there is relative movement between the magnet and coils, but they are centered, no current flows since the potential cancels out. When they are displaced off-center, current flows and a strong field is generated by the coil, which tends to restore the spacing. For levitation, the repulsive forces of the lower loops and attractive forces of the upper loop cause the vehicle to lift. The ratio of the lift and drag forces may be a measure of the viability of these systems in vehicle applications. The loss in ground coils should be reduced to increase the ratio of the lift and drag forces. Thus, the m.m.f. in ground coils must be. The ratio is defined by

Fig. 4.14 Combination of
rectangular superconducting
and ground coils

(a)

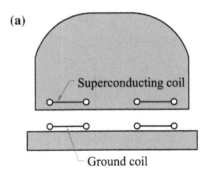

(b)

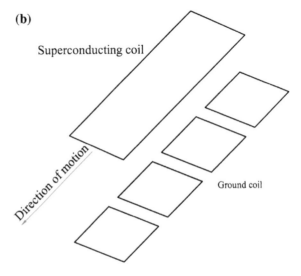

$$\text{Drag ratio} = \frac{vL}{R}\frac{\partial M}{\partial z}\frac{1}{M} \qquad (4.16)$$

where L and R are the inductance and resistivity of the ground coil, respectively,
and M is the mutual inductance between the superconducting and ground coils.

Equation (4.16) suggests that a lower M and larger derivative of M with respect
to z are desirable to increase the drag ratio. The null-flux system satisfies these
requirements, as shown in Fig. 4.17. Figure 4.18 shows the distributions of flux and
induced currents in the ground coils in a 8-shaped.

If the superconducting magnet is located in the center of the levitation coil, the
mutual inductance M is 0. M is proportional to the vertical displacement from the
center of the levitation coil in 8-shaped. There is an equilibrium position where
the vertical force from the field gradient is equal and opposite to the weight of the
suspended vehicle mass. The position is naturally located below the center of the
levitation coil. The equilibrium position is about 40 mm in the vertical direction.

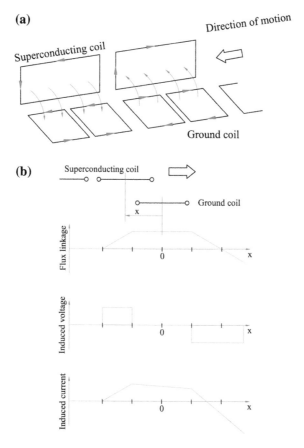

Fig. 4.15 Levitation principle of superconducting coils over perpendicular ground coils

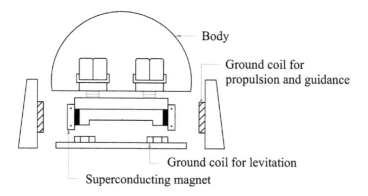

Fig. 4.16 System configuration of the MLU-001

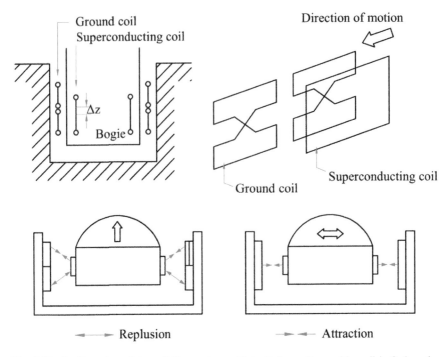

Fig. 4.17 Configuration of the null-flux system with levitation coils on side wall in 8-shaped

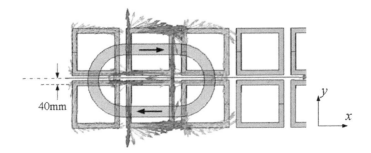

Fig. 4.18 Distribution of flux and induced currents in the ground coils [7]

For lateral guidance, the left and right coils are connected by a cable to form null-flux, providing centering forces. The restoring force is proportional to lateral displacement. This system does not require levitation and lateral guidance control, which is its main advantage. However, this system needs a cryostat for keeping the superconducting coil at below critical temperature. With the development of a high-temperature superconducting material, this system might be expected to be more viable.

4.6 Passive Damping

To compensate for the low damping inherent to the superconducting system, passive and active control methods may be introduced. In the passive method, an aluminum or copper plate acting as a shield is inserted between the magnet and the conductive sheet, the shield being rigidly attached to the magnet Fig. 4.19. Whenever there is vertical movement, the magnetic field created by the conductive sheet induces eddy currents in the aluminum or copper plate, which in turn add some damping force in the levitation system. The amount of damping force is dependent on the thickness of the damper plate and its distance (d) from the levitation magnet. To assess the effectiveness of the damper plate on damping force, damping time constant and damping ratio need to be considered. For a suspension height z, the damping time constant τ can be defined as

$$\tau = \frac{2w_p(2z - d)}{zg} \tag{4.17}$$

where w_p is the plate parameter and is expressed as $w_p = 2/\mu_0\sigma_p h_p$. The optimum value w_p for aluminum sheet at any frequency of oscillation f may be derived to be [6]

$$w_{popt} \cong 2.45\pi f(2z - d) \tag{4.18}$$

Combining the above two equations, the optimum damping ratio that can be obtained with a passive aluminum plate is given by

$$\xi = \frac{1}{2\pi f\tau} \text{ with } \tau \cong \frac{4.9\pi f(2z - d)^2}{zg} \tag{4.19}$$

Since d is positive and $\tau \geq 0$, Eq. (4.19) may be written as

$$\left(2 - \frac{d}{z}\right)^2 = \frac{\tau g}{4.9\pi fz} \tag{4.20}$$

which yields the following approximate bounds for $0 < d < z$:

$$6.28 < \frac{\tau}{fz} < 1.57 \tag{4.21}$$

For an intermediate value $\tau/fz = 4$ with $f = 1.6$ Hz and $z = 10$ cm, the damping time constant is 0.64, giving $d \cong 4$ cm and $h_p = 1.3$ cm at room temperature. It is known that the damping ratio attainable through this method is around 0.2. For vehicle suspension this damping ratio represents a significantly underdamped system, and additional damping will need to be incorporated either through a

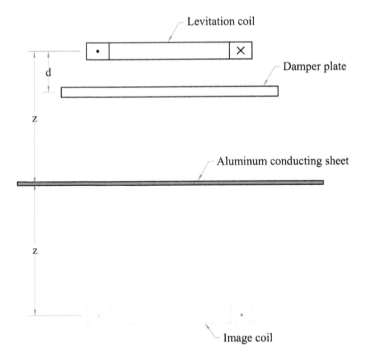

Fig. 4.19 Physical arrangement of achieving damping through passive plate

secondary suspension system or through an active controller described in the following section. Damping remains a outstanding problem associated with the
superconducting levitation systems.

4.7 Active Damping Control

Three methods of active damping control schemes are briefly introduced in this
section [6]. The descriptions of the methods aim at giving their main concepts and
features. The first method is the dynamic control of the current in the superconducting coil (magnet) (Fig. 4.20). In this scheme, the instantaneous current through
the superconducting coil is split into a steady-state component $[I_0]$ which provides
the equilibrium levitation force (mg), and a control component $[I_{cl}$ (t)] which
generates the transient damping force. The vertical equation of motion may be
described by extending the Eqs. (4.6) and (4.7)

$$m\ddot{z}(t) = -mg + F_l[I_0, i_{cl}(t), z(t)] \qquad (4.22)$$

$$F_{l0}(I_0, z_0) = mg \qquad (4.23)$$

and

$$m\Delta\ddot{z}(t) = \left.\frac{\partial F_l}{\partial i_{cl}}\right|_{(I_0,z_0)}\Delta i_{cl}(t) + \left.\frac{\partial F_l}{\partial z}\right|_{(I_0,z_0)}\Delta z(t) \qquad (4.24)$$

$$m\Delta\ddot{z}(t) = \beta_i\Delta i_{cl}(t) + \beta_z\Delta z(t)$$

where the vertical lift force, at any time t, is given by Eq. (4.2).

$$F_l[I_0, i_{cl}(t), z(t)] = [I_0 + i_{cl}(t)]^2\frac{dM(\hat{z}, t)}{d\hat{z}} \qquad (4.25)$$

If the transient current $i_{cl}(t)$ was derived by using the linear control law

$$i_{cl}(t) = -k_{pl}\Delta z(t) - k_{vl}\Delta\dot{z}(t) \qquad (4.26)$$

then the error dynamics of the closed-loop system are given by

$$\Delta\ddot{z}(t) + \frac{\beta_i k_{vl}}{m}\Delta\dot{z}(t) + \left[\frac{k_{pl} - \beta_z}{m}\right]\Delta z(t) = 0 \qquad (4.27)$$

which suggests that the desired damping may be achieved by controlling the velocity feedback gain. Furthermore, the position error feedback also gives the flexibility of controlling the natural frequency of the closed-loop system. This method is conceptually simple, but poses major difficulties if the superconducting coils are to be operated in persistent current modes.

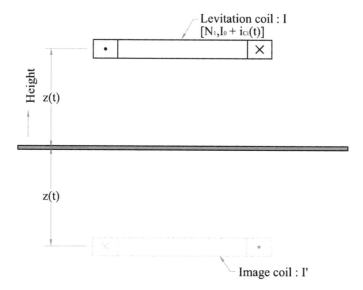

Fig. 4.20 Active damping control through combined levitation and control coil

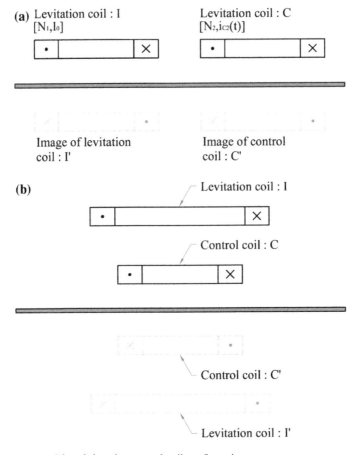

Fig. 4.21 Separate lift and damping control coil configurations

The second method is to use an additional coil for damping control. Two configurations may be used in this scheme, as shown in Fig. 4.21. The control law in Eq. (4.26) can be equally applied to this configuration through modifying the added forces among the levitation and control coils and the conductive plate. If the control coil is placed at room temperature, the control current requirement may be much higher than the control current with the first method. This considerably higher control current requirement in the second method (separate lift and control scheme) would need to be contrasted with the advantages of having two separate coils and driving the levitation coil in persistent mode.

In superconducting maglev vehicles in operation, the damping of the primary suspension, i.e. the superconducting magnet, may be negligible. To add damping force in the primary suspension for ride comfort, the introduction of actively controlled electromagnetic damping was proposed [8]. The configuration is illustrated in Fig. 4.22. The power is generated by the interaction between the on-board

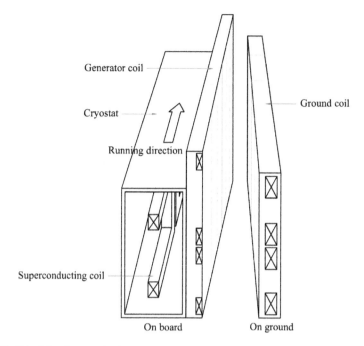

Fig. 4.22 Principle of magnetic damper using a linear generator [8]

linear generator coils and the magnetic fields produced by the ground coils. The basic principle of the scheme may be similar to the separate lift and control method described earlier. If the phase of the current in the power collecting coil is controlled, then the lateral and vertical damping force can be adjusted to suppress the amplitude of harmonic vibrations of the primary suspension.

4.8 Discussion

Magnetic levitation through superconducting magnets both in static and dynamic modes offers the advantage of levitation stability without any levitation control action. Moreover, the discovery of high-temperature superconducting materials may offer a promising option for transportation and conveyor systems. One of the outstanding problems is the inherently low damping. It can be said that the vibration control with some means is a subject in this area that is open for research and development.

References

1. Photo courtesy of CJ Kim at KAERI
2. Source: IFW Dresden, Germany, http://www.supratrans.de
3. Stephan R, David E, Andrade Jr., R, Machado O, Dias (2006) A full-scale module of the Maglev-Cobra HTS-superconducting vehicle. Maglev 2008, San Diego, USA
4. Dias DHN, Sotelo GG, Andrade Jr (2014) Dynamic and static measurements with the basic cryostat unit of a superconducting magnetically levitated vehicle. Maglev 2014, Rio de Janeiro, Brazil
5. Photo courtesy of COPPE/UFRJ
6. Sinha PK (1987) Electromagnetic suspension dynamics & control. Peter Pergrinus Ltd., London
7. Cho HW, Bang JS, Han HS, Sung HK, Kim DS, Kim BH (2008) Status of advanced technologies and domestic researches for development of Korean next generation Maglev. Korean Inst Electr Eng 57(10):1767–1776
8. Watanabe K, Suzuki E, Yoshioka H, Murai T, Kashiwago T, Tanaka M (2004) Vibration control of maglev vehicles utilizing a linear generator. Maglev 2004, Shanghai, China

Chapter 5
Electromagnet

5.1 Introduction

All the maglev trains currently in service or to be operated in several years employ
electromagnetic levitation systems, rather than permanent or superconducting
magnet systems. This includes Transrapid, Linimo, ECOBEE, and the other three
Chinese urban maglev trains. Furthermore, most of the attempts at using maglev to
transfer LCD glasses and semiconductors in an extremely clean environment are
also based on electromagnet systems, which are introduced in Chap. 7. As such, the
electromagnetic attraction systems of the magnetic levitation systems proposed are
most widely used in various applications, and thus most of the research on magnetic
levitation has been focused on this type of magnet. An electromagnet consists
simply of a ferromagnetic core, such as steel, and a current-carrying winding wound
on the core. While the manufacturing and operation of the magnet is relatively easy,
a sophisticated feedback control system needs to be incorporated to maintain a
constant separation between the pole face of the magnet and the ferromagnetic
reaction surface, as the system is inherently unstable due to the use of attractive
forces proportional to its separation. That is, if the constant airgap control does not
work appropriately, the two bodies will be attracted or diverged. For this reason, the
system must be stabilized through sophisticated feedback control. This is what
makes it different from the permanent and superconducting magnet systems mentioned in the preceding chapters. Although interdisciplinary research is needed for
the stabilization of electromagnets, the motion controllability of the electromagnetic
systems, even in high precision applications, is a primary advantage. In addition,
this system can be levitated at zero speed. Due to such primary features, electromagnetic systems have potential applications in diverse areas. This chapter presents
the basis for the design, analysis and implementation of the electromagnetic systems. The performance requirements, design considerations, magnet design

© Springer Science+Business Media Dordrecht 2016
H.-S. Han and D.-S. Kim, *Magnetic Levitation*,
Springer Tracts on Transportation and Traffic 13,
DOI 10.1007/978-94-017-7524-3_5

procedure, control scheme, control and measurement and electronics are briefly
introduced. The configuration, principles and features of the electromagnetic levitation systems are first given in Sect. 5.2. In Sect. 5.4, the design procedure and
considerations that must be applied to obtain the required attraction forces are
outlined. A considerable part of the chapter, Sect. 5.5, is devoted to the feedback
control schemes needed for stable suspension in the electromagnetic systems. As
the basis for the control schemes, the mathematical models for these levitation
systems are examined, as well as the related assumptions. In Sect. 5.6, the passive
and active guidance control schemes in the lateral direction are briefly introduced.
One of the main problems in the electromagnet systems that significantly influences
the economic and stability aspects is the dynamic interaction between a vehicle and
an elevated guideway. The dynamic interaction problem is reviewed in Sect. 5.7.
Brake and switch needed for the operation of the electromagnet systems are also
surveyed in Sect. 5.8. To help the reader understand the contents in the Chapter, an
experimental vehicle SUMA550 in Sect. 5.3, as a case study, is used in Sects. 5.4,
5.5 and 5.6.

5.2 Levitation

5.2.1 Principle

The combination of a U-core electromagnet and a ferromagnetic reaction surface
(track) is shown in Fig. 5.1 to illustrate the principle of electromagnetic levitation.
Under this principle, the term levitation, as used in this Chapter, means that the

Fig. 5.1 Electromagnet-track
configuration

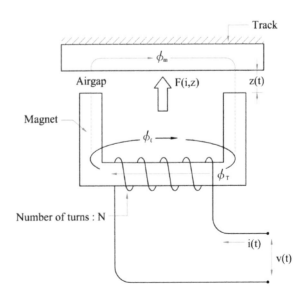

magnet that is free to move is suspended continuously at a distance from the fixed track. For a vehicle suspension system, the earlier arrangements in the repulsive modes of the permanent and superconducting magnets will have to be inverted in the electromagnet systems. For this reason, though the term "suspension" may be more appropriate for the electromagnet systems, the term "levitation" is used in this chapter for consistency in the context of magnetic levitation in the monograph. An electromagnet simply consists of a ferromagnetic core, such as steel, and a current-carrying winding wound on the core. The basic properties of an electromagnet were already given in Chap. 2. The strength of the magnetic fields generated by an electromagnet is proportional to the currents and the number of turns in the coil. The magnetic flux from the core forms a flux path through the track with a higher permeability than air. If the flux path is closed, the force of attraction between two bodies is produced, the magnet being attracted towards the track. If the current in the magnetic coil is controlled to maintain a constant clearance (airgap), the electromagnet can be suspended by the controlled attraction force, without any mechanical connection. This is the principle of levitation through electromagnets. If this concept is applied to a vehicle, the configuration is the same as that shown in Fig. 5.2, in which the lift magnets are placed in the vehicle and the ferromagnetic track is mounted underneath the guideway. While maintaining the clearance by adjusting the current in the magnetic coil, if the vehicle is propelled by a linear

Fig. 5.2 Configuration of a train using the attractive electromagnetic levitation

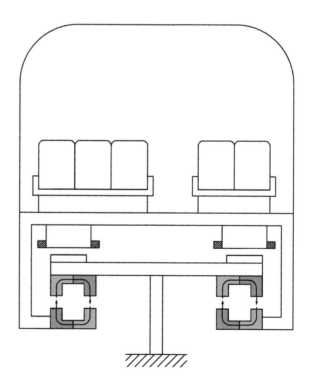

Fig. 5.3 Suspension of an
object by an electromagnet
fixed to ground

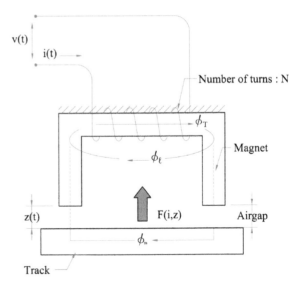

motor, it becomes a maglev vehicle running without wheels. It is worth noting here
that the magnets are located underneath the track, unlike the examples of the
permanent and superconducting magnet systems. This configuration indicates that
the feedback control loop for maintaining a constant clearance is the most critical
element for the performance of this kind of maglev vehicles. Furthermore, it suggests that some safety gears are needed for safe operation in the event that the
levitation systems with the controlled electromagnets function abnormally.

The physical arrangement in Fig. 5.1 can be also inverted to be as shown in
Fig. 5.3, with the magnet fixed to ground and the ferromagnetic object suspended.
Using the same constant airgap control scheme as for the system in Fig. 5.1, the
object can be suspended at a distance from the electromagnet. If a linear motor is
employed to propel the suspended object without physical contact, the object can
move no generating particles, especially in an extremely clean room. The main
feature of the configuration in Fig. 5.4 is that there is no need for a power supply for

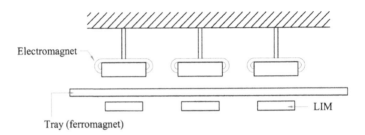

Fig. 5.4 Configuration of the conveyor proposed for the steel tray carrying LCD glasses

the levitated object made of ferromagnetic materials. On the other hand, one of the limitations in this system is the large number of electromagnets mounted throughout the whole length of the track, each with its drive and position sensor. This concept could be used for conveying LCD glasses in an extremely clean environment. The application of the concept will be outlined in Chap. 7.

5.2.2 Properties

It is relevant here to consider the nature of the electromagnetic attraction system described in the preceding section, because this type is significantly different from both the permanent and superconducting magnets. For convenience, the expression for the attraction force between two magnetized bodies is first used here for relating the parameters contributing the force, though it will be taken into account more in the following sections. The attractive force $F \propto (i(t)/z(t))^2$. That is, if the current is assumed to be constant, then the force increases as the airgap decreases, and eventually two bodies will be attached to the ferromagnetic object (track). Inversely, the force will be decreased as the airgap increases, resulting in the separation of the two objects. These inverse force-distance characteristics are illustrated in Fig. 5.5. Consequently, if the current is assumed to be constant this system is inherently unstable, and will not maintain a constant airgap. That is, if any external force disturbance enters the system in an equilibrium position, the two objects would either be adhered or separated. This property suggests that an active control means should be incorporated in the system for stable suspension, adjusting the currents in the magnet coil with the clearances. For the stabilization of the attraction type systems, the function of the control system should modify the force-distance characteristics of the electromagnet such that the slope of the attraction force with respect to the distance is positive (Fig. 5.6) [1]. One of

Fig. 5.5 Force-distance characteristics of an electromagnet with constant current

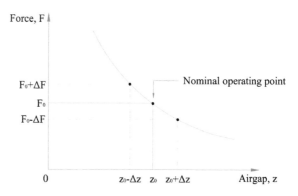

Fig. 5.6 Stabilization
through the position feedback:
D.C. explicit method

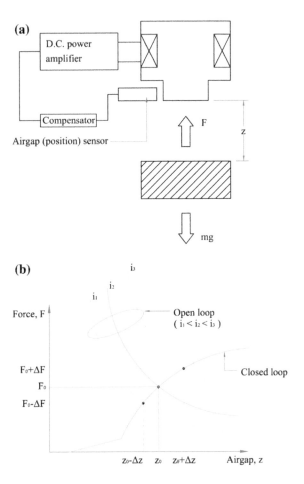

the methods for modifying the force-distance characteristics is the explicit method
using the external sensor to measure the distance or clearance ("airgap" hereinafter).
The strength of the magnet is compensated for through the position feedback loop
controlling the magnet coil current (Fig. 5.6a). The position error feedback loop is
designed in such a manner that the electromagnetic attraction force is proportional
to the airgap around the equilibrium position (Fig. 5.6b).

The features of these electromagnetic levitation systems can be summarized as
follows.

• Due to its inherent instability, the system should be stabilized through an active
  control action.
• The airgap is very small compared with the permanent and superconducting
  systems, due to considerations for power consumption and operation within

linear range. Appropriate airgaps are around 10 mm for vehicles, and several mm for conveyor applications.

- As the system has the capacity to adjust the nominal position and restrict the variation in the position, it is capable of controlling the position, even with high precision. As such, it could also be applied to high precision instrumentations.
- Using the control law to derive the current in the magnet coil, the levitation stiffness and damping are controllable. This feature enables an acceptable ride comfort to be achieved in passenger carrying vehicles. As was mentioned earlier, both the permanent and superconducting systems have very low damping, and furthermore, it is difficult to add damping in the systems.
- This system can operate at room temperature, i.e. there is no need for a cryostat.
- The assembly and maintenance is easier because the force is produced when the current flows.
- Levitation at zero speed, i.e. stationary levitation, is possible.
- A favorable power supply is needed due to its smaller airgaps.
- The implementation of the electromagnetic system may be economically feasible with the development of lower-cost measurement and power electronics devices.
- In contrast, a sophisticated levitation control system is required due to its inherent instability. To implement the control system, electronics, measurement and control theory, mechanics, and expertise in other disciplines is necessary.

A comparison of the three types of magnets is provided, in Table 5.1. The information provided in this table is cited from Ref. [1], with few modifications.

5.2.3 Performance Requirements

If electromagnets are used as levitation systems in any form, the performance requirements may be similar to those of the primary suspensions of conventional road and railway vehicles, which are rubber tires or steel wheels. Furthermore, all of the primary vehicle suspensions, in any form, can be simplified into a simple linear or non-linear spring-damper system. Usually, a linear model is in first analysis widely used for design and analysis. The key parameters for the suspension systems, from a dynamic viewpoint, are the stiffness and damping, which give the natural frequency and damping ratio of the system. The most important task in designing a magnetic suspension system with electromagnets is to obtain the desired stiffness and damping properties by adjusting its physical properties and control schemes. In passenger carrying vehicles, adequate stiffness and damping are dictated by the passenger comfort and the track holding properties. The function of the natural frequency and damping ratio in the electromagnetic systems is equally applied to the permanent as well as the superconducting systems; the only difference is that the former is active suspension and the latter is passive. In general, the

Table 5.1 Comparison of three types of levitation magnets

Magnet type	Advantages	Limitations
Electromagnet	Stable suspension with stationary magnets	Large I^2R loss
	Inexpensive guideway material (steel)	Nonlinear and eddy-current effects
	Uses established control techniques and commercially available components	Thermal rating limits magnetic fields
Superconducting magnet	Low I^2R loss allows large magnet currents and high magnetic fields	Very low intrinsic damping
	Cost of guideway is comparable with modern high-speed wheeled system	Requires auxiliary suspension below critical speeds
	Smaller conductors are needed for higher current density	Additional components for on-board refrigerator/cryostats
	Large clearance makes the vehicle less sensitive to guideway irregularity	Comparatively expensive guideway material (aluminum)
Permanent magnet	For large volume productions, magnets are no more expensive than steel	Static repulsion or attraction fields require a mechanical contact for stability
	Continuous supply of power is not needed	Susceptible to loss of field due to heating
	Very simple configuration and lower maintenance cost	Because of the low field strength, contact/low clearance is needed

advantages and limitations of passive and active systems would apply equally to electromagnetic suspension systems. The practical meaning of the two parameters for an electromagnetic system is relevant for its design and analysis.

- **Suspension stability**: Under all operating conditions and disturbances, the electromagnetic system should be capable of maintaining its suspension. This ability is the primary performance requirement. For passenger carrying vehicles, there are four sources of excitation: (1) suspended load changes, (2) unsteady aerodynamic forces, (3) guideway induced vibrations, and (4) guideway roughness and misalignment. Even though the relative influences are different based on the particular applications, most disturbances can be categorized into one of these four sources. In addition, an operational scenario such as a velocity profile may be added. Prior to the system design, the static and dynamic design requirements must be well established after considering all the operating conditions and disturbances above. Consequently, a sufficient stability margin should be provided with the electromagnetic systems. From a broader viewpoint of stability, though the following required performances belong to the suspension stability described above, for design convenience, they are slightly more addressed below.

- **Track holding**: The typical elevated guideway (track) has some profile irregularities. These irregularities are due to the geometrically uneven surface and the static and dynamic deflections of the guideway when a vehicle moves over it. They are characterized as relatively large amplitude waves at low frequencies, and as small amplitude waves at higher frequencies. For example, guideway deflection belongs to the former, while surface roughness is considered as the latter. One of the primary performance requirements of a levitation system is a stable suspension under all operating conditions, with acceptable acceleration levels. To follow the guideway profile well, stiffer suspension is needed. In contrast, reducing the vertical acceleration level requires a softer suspension, to improve ride comfort. Consequently, the suspension stiffness design is based on a compromise between the acceleration level and the track (guideway) holding property. To achieve this compromise, the amplitudes and frequencies of the guideway irregularities under consideration must first be identified. That is, some of them are strictly followed and others are ignored. The levitation control scheme would be refined considering the classifications.

- **Stiffness**: (k_s): In magnetic levitation, stiffness is defined as the change in airgap per unit change in levitated load. This stiffness is related to a steady-state error in the position control loop of the electromagnetic systems. The variation in airgap means the track holding property. Because the ratio of lift force to input power reduces with an increasing airgap, the stiffness that determines the magnitude of variation in airgap plays an important role in the design of magnets for obtaining lift capacity and the selection of a suitable active control system. Stiffness may be increased by increasing the loop gain or by incorporating an error integral feedback into the levitation system. The loop gain is increased by increasing the magnet's lift capacity, the rated voltage, and the power amplifier's gain. However, there is an upper limit to the extent to which this can be increased, because noise and vibration would be induced by the saturation effect in the magnet and the stronger coupling with the guideway. In the error integral feedback action technique, the steady-state errors may be reduced, but the transient responses may likely be deteriorated.

- **Natural frequency**: The mass of the magnet and its associated components and the stiffness described earlier provide the natural frequency of the system. As indicated earlier, the stiffness of the levitation control loop is chosen based on the control law and physical properties. The natural frequency of the control system needs to be sufficiently separated from the guideway's fundamental bending frequency and the vehicle pylon passing frequency proportional to vehicle speed. Since a lightweight guideway is usually preferred for cost reasons, the guideway's bending frequency may be likely to become closer to the magnet's natural frequency of around 10 Hz. In addition, for vehicles, the structural vibration modes of the bogie carrying the magnets are likely to have natural frequencies closer to the magnet's frequency. For this reason, it is

recommended to ensure that no resonances can occur with any possible excitation forces by separating the vibration modes.

- **Damping**: Damping in the electromagnetic systems is critical for suspension
  stabilization. The acceleration levels, a measure of ride comfort, are mainly
  limited by the damping. Damping can be adjusted based on the velocity feedback scheme in the levitation control system. In general, a damping ratio of 0.5–
  0.8 is required in passenger carrying vehicles.
- **Ride quality**: The ride comfort is influenced mainly by the secondary suspension in vehicles with a natural frequency of around 1 Hz and a damping ratio
  of around 0.7. However, because the primary suspension, the magnet, also
  affects ride comfort, the equivalent stiffness and damping of the magnets are
  chosen when assessing the acceleration levels.
- **Power requirements**: After determining the nominal airgap and its variation,
  the static and dynamic power requirements are chosen based on the stiffness.
- **Stray magnetic field**: There may be a guideline to limit the stray magnetic field
  strength produced by the electromagnets. The stray magnetic field strengths
  need to be measured or predicted to ensure there is no risk to human health.

5.2.4 General Configuration

Figure 5.6a conceptually shows the configuration of a controlled electromagnetic
attraction system with only a position sensor. As indicated earlier, the fundamental
requirement of the maglev system is its capacity to maintain its levitation under all
operating conditions and external disturbances. The deflection of the guideway with
large amplitude may create disturbances equivalent to 100 % variations in
airgap. Furthermore, since the nature of a small airgap and its variation makes the
system strongly coupled, the design of the system requires that more careful consideration be given to static and dynamic analysis than in the passive permanent and
superconducting systems. As such, a more generalized control scheme may be
needed in practical systems. The relative or absolute position, velocity, and
acceleration as well as flux may be selectively used in the feedback loops for stable
levitation. One of the possible configurations, which is now being used, is given in
Fig. 5.7, and uses a position sensor and an accelerometer.

The configuration consists of an airgap sensor, accelerometer, electromagnet,
ferromagnetic guideway, power amplifier and power source. The function of each
component can be summarized as follows.

- **Gap sensor**: This measures the distance between the pole face and the guideway
  surface.
- **Accelerometer**: The acceleration, usually in the vertical direction, is measured,
  and used to derive velocity and position by integrating it.
- **Electromagnet**: The electromagnet is the actuator for levitation. It is desirable
  to minimize its time constant.

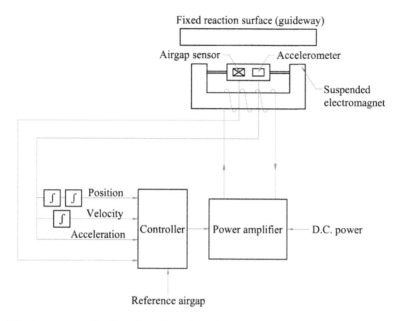

Fig. **5.7** A more generalized configuration of the electromagnetic levitation system

- **Guideway**: The guideway is usually made of ferromagnetic materials such as steel. Because in maglev systems the guideway is supported by piers at a distance of 20–40 m, the deflection of the guideway due to a vehicle moving over it is not negligible.
- **Controller**: The levitation controller performs the functions of signal processing and derives the excitation current in the magnet coil by the control law.
- **Power amplifier**: This is virtually a driver for a magnet consisting of transistors.
- **Power source**: This supplies the required electrical power for the power amplifier.

A maglev vehicle based on the generalized configuration described above may become the system shown in Fig. 5.8, which is very similar to the one used for Transrapid. The details of this system will be considered later. In addition, this concept or variations of it could also be applied to other applications.

The possible configurations of using the controlled electromagnets for levitation and guidance are illustrated in Fig. 5.9. All maglev vehicles currently in operation use one of the configurations provided in Fig. 5.9. These configurations can be classified into a combined lift and guidance and a separate lift and guidance. The former provides both lift and guidance forces with one magnet. As only one reaction surface is needed, the construction cost of the guideway would be considerably reduced. On the other hand, this approach imposes limitations on the magnet-guideway geometries and requires a rather complex control scheme because of its cross-coupling. The main advantage of the separate lift and guidance system is the orthogonality of the

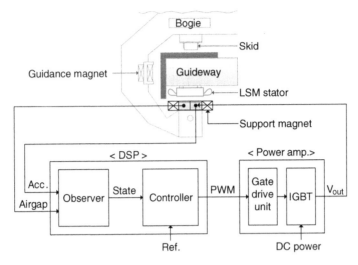

Fig. 5.8 Application of the more generalized configuration to a maglev vehicle (similar to Transrapid)

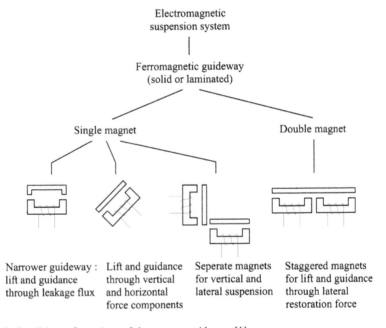

Fig. 5.9 Possible configurations of the magnet-guideway [1]

two magnetic planes, leading to independent control of the vertical lift and lateral guidance forces. Its main limitation is the need for an additional reaction surface for the guidance magnet. For higher speed vehicles, separate magnets are preferred due to the large centrifugal forces during curve negotiation. The combined configuration is being used for low-speed vehicles for urban transit, in which the guidance force is less than about 30 % of the vertical lift force.

5.3 Experimental Vehicle

An experimental vehicle was constructed to study the design and analysis of the electromagnets for lift and guidance, as well as their corresponding control loops [2] (Fig. 5.10). A research team with members from KRRI (Korea Railroad Research Institute), KIMM (Korea Institute of Machinery and Materials), Woojin Industrial Systems and KR (Korea Rail Network Authority) was formed and funded by the KAIA (Korea Agency for Infrastructure Technology Advancement). The vehicle has the separate support and guidance magnets referred to earlier, the support magnets being also used as a field system for the long-stator LSM installed on the guideway (Fig. 5.11). The properties and design parameters of the vehicle are listed in Table 5.2. The vehicle has 6 electromagnets for levitation, 4 electromagnets for guidance, and 2 electromagnets for braking. The nominal airgaps for levitation and guidance chosen were identical, at 10 mm. The lift magnet, which

Table 5.2 Experimental vehicle specifications [2]

Parameters	Values
Max. design speed	550 km/h
Length	13 m
Suspension	Electromagnet
Guidance	Electromagnet
Braking	LSM/Electromagnet
Propulsion	Iron-cored long-stator LSM
Suspension airgap	10 mm (operating)
	20 mm (landing)
Guidance airgap	10 mm
Acceleration/Deceleration	1.1 m/s^2
Car body mass	12,576 kg
Total bogie mass	10,500 kg
Air spring stiffness	73,689 N/m
Damping coefficient	15,600 Ns/m
Number of lift magnets	6
Number of emergency braking magnets	2
Number of guidance magnets	4

Fig. 5.10 General view of an experimental maglev vehicle with the electromagnets for suspension and guidance [2]

served as the field system for LSM, provides a sufficient magnetic flux density for a speed of 550 km/h. The secondary suspension consists of conventional air springs with some damping. The test track length is 150 m, allowing a maximum attainable speed of up to 30 km/h (Fig. 5.10).

5.4 Magnet Design

5.4.1 Procedure

As indicated earlier, the fundamental function of the electromagnet is to provide stable suspension for the vehicle under all operating conditions and disturbances. In addition to the primary function, the short-term and long-term economic benefits should be considered. To meet those requirements, the magnet design is influenced by the dynamic and static characteristics. The static characteristics include the electrical power requirement, power loss in the magnet, weight of magnet and magnet-guideway geometry. Moreover, the magnet should be capable of providing a sufficient margin of stability without an excessive increase in magnet weight per power input. The static and dynamic requirements can be satisfied through the well-established magnet design procedure [1] (Fig. 5.12). During the design, electromagnetic field simulation software can be usefully applied. The static, frequency-domain, and time-varying electromagnetic and electric fields can be analyzed in 2D/3D using the commercial magnetic field simulation software. In choosing the appropriate design variables and parameters, the complementary use of a simplified and rigorous analysis model is recommended. Although general purpose programs can be usefully applied to analyze the static and dynamic characteristics, the simplified analysis model can be still employed in the first analysis. As such, the design procedure with a simplified analytical model is outlined in the following sections.

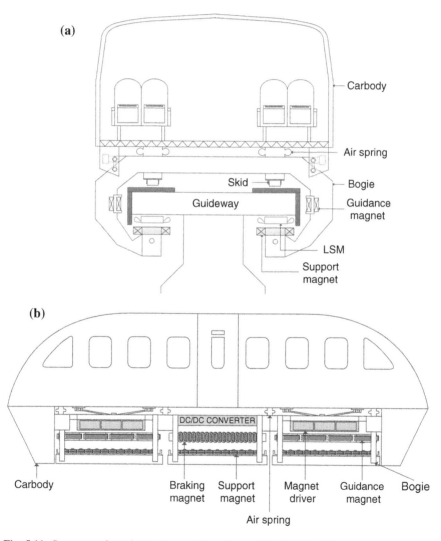

Fig. 5.11 System configuration: **a** cross-section view and **b** side view [2]

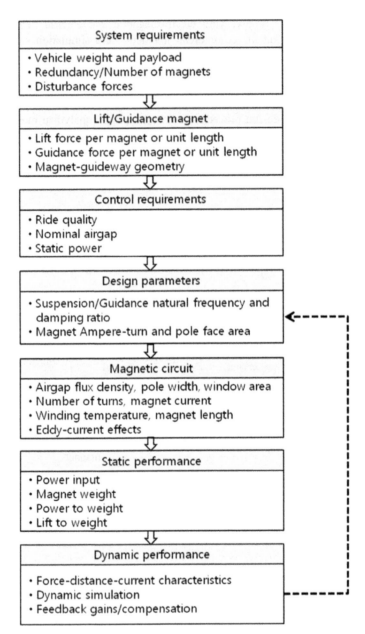

Fig. 5.12 Outline flow-chart for magnet design [1]

5.4.2   *Attraction Force*

The force of attraction generated by the electromagnet depends on the geometries of magnet and ferromagnetic object (guideway). As such, the selection of their basic geometries would be much more important to achieve the required lift force and construction costs. In addition, since most of the maglev systems need some guidance forces in practical applications, an adequate magnet guideway combination needs to be chosen in order to obtain the levitation and guidance forces. The possible magnet guideway combinations are given in Fig. 5.13 [1]. U-core magnet and E-core magnet are the representative ones. The combination of an U-core magnet and a guideway provides the vertical lift force as well as some horizontal guidance force by the fringe flux. The guideway can be made in a flat or a trough shape, and the trough provides more guidance force than a flat one. When separate lift and guidance magnets are used, the dimensions of a magnet are chosen to

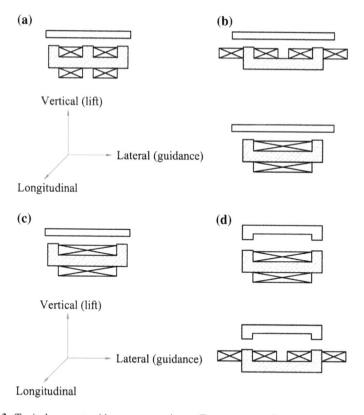

Fig. 5.13 Typical magnet-guideway geometries: **a** E-core magnet, **b** U-core magnet, **c** U-core magent with flat track, and **d** U-core magnet with trough track

Fig. 5.14 Dimensions of a
U-core magnet: **a** core,
b winding

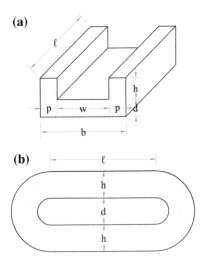

provide the pole face area (*pl* in Fig. 5.14) required to obtain sufficient lift forces
and to offer the needed window area (*wh* in Fig. 5.14) to house the excitation coils
[1]. A U-core configuration needs a relatively wider guideway for a sufficient
window area. On the other hand, the eddy currents induced in the guideway have
been found to be substantially less than E-core magnets. For these reasons, U-core
magnets are widely used for low- and high-speed vehicles. Thus, the design and
operational features of U-core magnets are briefly presented in the following
sections.

- **Pole face area and window area**: The force of attraction between two
  high-permeable surfaces in Fig. 5.15 may be defined as

$$F = \frac{dW_m}{dz} = -\frac{1}{2}\phi\frac{dM}{dz} \qquad (5.1)$$

where W_m is magnetic co-energy, M is the airgap m.m.f. and the flux ϕ linking
the two surfaces is assumed to be invariant with respect to z. If λ represents the

Fig. 5.15 Flux density
distribution between two
magnetized bodies

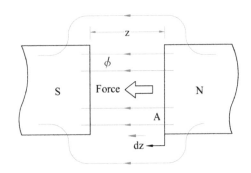

permeance of the airgap, with the assumption of the negligible reluctances of the two high-permeable media, then

$$\lambda = \frac{\mu_0 A}{z} \quad \text{and} \quad M = \frac{\phi}{z} \tag{5.2}$$

Rearranging the above equations

$$F = -\frac{1}{2}M^2 \frac{\mu_0 A}{z^2} \equiv \frac{B^2}{2\mu_0}A \tag{5.3}$$

where

$$B = \mu_0 H = \frac{\mu_0 M}{z} \rightarrow \text{airgap flux density}$$

For the U-core magnet in Fig. 5.14 with two pole face areas, the above force equations give rise to the attraction force expression

$$F = \frac{B^2}{2\mu_0}(2pl) \tag{5.4}$$

One of the two bodies is fixed to the ground, and the attractive force produced by the flux will make the free body move towards the fixed body (guideway). To relate Eq. (5.4) to the window area, the influence of the dimensions of the magnet on the attraction force needs to be identified. By deriving the force per unit area of the magnet from Fig. 5.14, the relationship may be obtained as follows.

$$
\begin{aligned}
F_a &= \frac{F}{(2p+w)l} \\
&\equiv \frac{B^2}{2\mu_0}\left[\frac{1}{1+\frac{w}{2p}}\right]
\end{aligned}
\tag{5.5}
$$

This relationship indicates that for constant B, the increasing $w/2p$ reduces the attraction force. Thus if the pole width is given, the window area may be determined by using Eq. (5.5). The next stage of design is concerned with temperature. If the whole window area is used to house the excitation coil, then choosing the coil width w is closely related to the temperature at the surface of the coil, as well as within the coil. If h is the height of either side and N is the number of turns in the coil, then the current density and thermal power density of a conductor carrying the current I with a resistivity ϱ are defined as [1]:

Conductor current density,

$$J = \frac{NI}{k_y k_z wh} \tag{5.6}$$

Thermal power density,

$$Q = k_y k_z \varrho J^2 \tag{5.7}$$

Winding packing (space) factors

Along y-axis:

$$k_y = \frac{\frac{1}{c_y} - \frac{1}{c_i}}{\frac{1}{c_c} - \frac{1}{c_i}} \tag{5.8}$$

Along z-axis:

$$k_z = \frac{\frac{1}{c_z} - \frac{1}{c_i}}{\frac{1}{c_c} - \frac{1}{c_i}} \tag{5.9}$$

where c_y and c_z denote the thermal conductivities along the y and z axis, respectively, and c_c and c_i represent the thermal conductivities of the conductor and interlayer insulation, respectively. The temperature $T(y, z)$ at any position within the coil can be obtained by solving the following Poisson's equation.

$$\frac{\delta}{\delta y}\left[c_y \frac{\delta T(y,z)}{\delta y}\right] + \frac{\delta}{\delta z}\left[c_z \frac{\delta T(y,z)}{\delta z}\right] = -Q = -k_y k_z \varrho J^2 \tag{5.10}$$

Equation (5.10) is a nonlinear equation because the coefficients c_y, c_z and ϱ depend on the temperature. Thus, the solution of Eq. (5.10) can be derived through numerical techniques. The approximate solution of Eq. (5.10) may be obtained using the assumptions that the coefficients and the rectangular surfaces ($\pm y$ and $\pm z$) temperatures are constant. Based on the assumptions, the temperature at any surface of any concentric rectangle, shown by the dotted lines in Fig. 5.16, is expressed as

$$4zc_y \frac{dT}{dy} + 4yc_z \frac{dT}{dz} = -4Qyz \tag{5.11}$$

where

$$z = \frac{2h}{w} y \rightarrow \frac{dT}{dz} = \frac{w}{2h} \frac{dT}{dy} \tag{5.12}$$

Fig. 5.16 Configurations:
a U-core and single coil,
b single coil

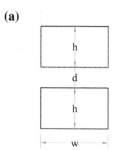

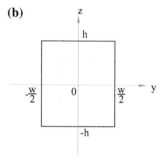

Substituting Eqs. (5.11) and (5.12) into Eq. (5.10), the Poisson's equation may be rewritten as

$$\left[c_y \frac{2h}{w} + c_z \frac{w}{2h} \right] \frac{dT}{dy} = -k_y k_z \varrho J^2 \frac{2h}{w} y \tag{5.13}$$

The temperature at the surface of the winding can be derived by integrating the above first-order differential equation over $y = 0$ (origin) and $y = w/w$ (surface).

$$\left[c_y \frac{2h}{w} + c_z \frac{w}{2h} \right] [T_{origin} - T_{surface}] = \frac{1}{4} k_y k_z \varrho J^2 hw$$

$$T_{origin} = T_{surface} + \frac{k_y k_z \varrho J^2 hw}{4 \{ c_y \frac{2h}{w} + c_z \frac{w}{2h} \}} \tag{5.14}$$

Since hw is the window area,

$$\text{Ampere-turns} = NI = k_y k_z Jhw \tag{5.15}$$

and consequently

$$T_{origin} = T_{surface} + \frac{\varrho (NI)^2}{4khw \{ c_y \frac{2h}{w} + c_z \frac{w}{2h} \}} \tag{5.16}$$

Fig. 5.17 Axial view of
U-core magnet

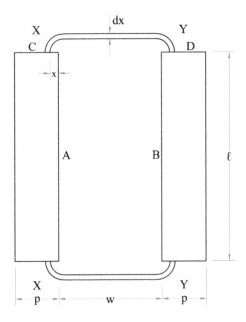

where $k = k_y k_z$ represents the overall packing factor of the winding, the current
rating of the winding may be determined from the specified maximum
temperature at the center of the coil, with a given ambient temperature ($T_{surface}$).
The simplified equation Eq. (5.16) may be only used for an initial estimate.
A more rigorous thermal analysis would be carried out by using the commercial
software to evaluate the most appropriate values of w and h for a given pole-face
area. The value of the maximum permissible temperature within the coil
depends on the coil geometry as well as the type of interlayer insulation. One of
the choices is aluminum coils wound with anodized foil, due to the absence of
airspaces in between layers and the resulting improvement in packing factor as
well as heat conductivity.

- **Airgap flux density**: To derive the attraction force from Eqs. (5.4) and (5.5), the
  airgap flux density B should be first chosen. It is, however, difficult to calculate
  this accurately because of the leakage flux. As indicated above, any electromagnetic field analysis program based on a 2D/3D model may be used with less
  simplification. For the electromagnet shown in Fig. 5.17 with $l \gg w$, the
  resulting equations may be usefully used to estimate the airgap flux density [1].

$$B = \frac{kwhJ}{pole\ area\ \times\ magnetic\ reluctance} = \frac{AT}{plR_m}$$
$$= \frac{kwhJ}{\left[2z + \frac{\mu_0}{\mu_i}\{2h + (1+\alpha)b\}\right]} \qquad (5.17)$$

where

$k = k_y k_z$ = composite packing factor of the winding
AT = ampere-turns
R_m = sum of reluctances in the three different media, i.e. core, airgap and guideway
$\mu_i = \mu_c = \mu_g$ = permeances of core and guideway
z = airgap
$\alpha = p/t_g$
t_g = guideway thickness
b = total width of magnet = $2p + w$
h = pole height

The assumptions and simplifications used in the derivation of Eq. (5.17) are not covered here. The reference material [1] is recommended for more details of the derivation.

- **Force, weight and power**: The basic performance index of a magnet is naturally its force capability and input power requirements. The two parameters that are widely used to quantify these are the lift force to magnet (core and coil) weight and lift force to input power (R_{lp}). Combining Eqs. (5.5) and (5.7), the attraction force between the magnet and the ferromagnet reaction surface, up to saturation, is given by

$$F = \frac{\mu_0 pl(kwhJ)^2}{\left[2z + \frac{\mu_0}{\mu_i}\{2h + (1+\alpha)b\}\right]^2} \tag{5.18}$$

The maximum available attraction force F_{max} is given by

$$F_{max} = \frac{B_{max}^2}{2\mu_0}(2pl) \tag{5.19}$$

The maximum airgap flux B_{max} is related to the saturation B_{sat} by $B_{max} = (\lambda_2/(\lambda_1 + \lambda_2))B_{sat}$. λ_1 and λ_2 are permeances of the main magnetic circuit and the leakage paths. With Eq. (5.18), the lift force to magnet weight ratio is given by

$$R_{lw} = \frac{F}{W_i + W_w} \tag{5.20}$$

where W_i and W_w are the weights of the magnet core and excitation winding. Next, the lift force to input power (R_{lp}) is defined by

$$R_{lP} = \frac{F}{P_0} \tag{5.21}$$

where $P_0 = I^2 r$ is the copper loss in the excitation winding. If ϱ is the specific resistance of the wire (copper or aluminum), then the resistance of the winding is given by

$$r = \varrho \frac{mean\ length\ of\ the\ winding}{winding\ area}$$

$$= \varrho \frac{2l + \pi(d+h)}{kwh}$$

Thus the copper loss in the excitation winding, with current $I(= kwhJ)$, is

$$P_0 = I^2 r = \varrho \left[\frac{2l + \pi(d+h)}{kwh} \right] (kwhJ)^2$$

$$= \varrho kwh \{2l + \pi(d+h)\} J^2 \qquad (5.22)$$

These parameters based on analytical derivation may be used for initial estimates.

5.4.3 Design Example

The interactive design procedure for the magnet design described in the preceding section has been well established and successfully used in the electrical machine area. To demonstrate the design of a magnet, the design of support and guidance magnets for SUMA550 in Sect. 5.3 is carried out as an example [2]. However, during the design works, the established design procedure and equations are not followed exactly. In addition, the performances of the magnets are assessed using electromagnetic field analysis software. The emphasis in this monograph is on the levitation and its control.

- **Combined lift and propulsion magnet design**: The lift magnet of the experimental and exemplary vehicle SUMA550 is shown in Fig. 5.18, which has two basic functions. One is to provide the levitation force. Another is the role of the field system for LSM for propulsion. Thus the reaction surface facing the magnet pole is not even because of the iron-cored stator configuration. Due to the iron-core stator profile, the airgaps are varied with respect to the magnet's position. Consequently, the levitation forces are influenced by the magnet's position relative to LSM stator and current in the LSM winding, resulting in vertical oscillatory motion. This motion is one of problems in the combined lift and propulsion magnet configurations with the iron-cored LSM. In designing the magnet, the following aspects were considered.

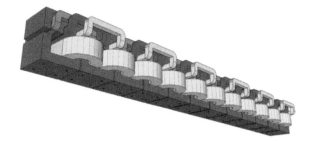

Fig. 5.18 Configuration of the lift magnet of the experimental vehicle SUMA550 [2]

- taking-off from 20 mm airgap of landing
- sufficient lift force and thrust
- variation in lift force due to varying current in the LSM winding
- aluminum sheet coil for winding for lightweight and heat dissipation

Considering the above design considerations, vehicle configuration, and specifications, the preliminary lift electromagnet is designed as shown in Table 5.3 and Fig. 5.19. The supporting magnet module has 12 poles, including two auxiliary poles at the ends. Allowing for the full load of 29,000 kg and a 50 % margin, the required lift force per unit length (1 m) will be 17 kN/m. The two auxiliary poles are for reducing fringing effects at the ends and ensuring space for the installation of gap sensors. The reduced detent force aspect was essential in the design, and is expected to lower the variations in lift force. Although there are various ways to minimize detent force, such as skewing the pole and pole pitch arrangement of the lift electromagnet, the latter technique is applied in the study because of the need for space for gap sensors, as well as to avoid difficulties in manufacturing. To minimize the net detent force, poles of lift magnet are arranged with intentionally non-synchronized positions. The result of this subdividing effect is that the more the module is subdivided into magnets, the less the detent force amplitude will be. This work proposes to reduce detent force through the arrangement of one pole per core instead of 12 poles. Figure 5.19c gives the pole pitch adjustment design with 12 individual magnets. Here, the central component of 12 magnets is synchronized with the corresponding LSM stator pole, and an interval of 7 mm is maintained between magnets. Figure 5.20 compares the detent forces in both designs. It shows that the detent force of the proposed design is less than 10 % of the preliminary one. The suspension force variations are also reduced as shown in Fig. 5.21, which would have positive effects on suspension control due to reduced lift force variations. Consequently, the reduction in detent force would lead to less vibration between vehicle and guideway. The lift forces are plotted in Fig. 5.22 with varying current in the magnet at two different airgaps. The rated lift force is achieved at 17 A at a nominal operating gap of 10 mm. It is noted that the employed design could, to some extent, lower thrust efficiency.

Table 5.3 Main design specifications of the lift magnet for SUMA550 [2]

Parameter	Value	Parameter	Value
Pole pitch	240 mm	Stator height	130 mm
Number of poles	12	Slot width	40 mm
Slot depth	50 mm	Tooth width	40 mm
Required lift force per unit length	17 kN/m	Required thrust	5 kN
Magnet width	220 mm	Magnet height	150 mm

Fig. 5.19 Designs of
electromagnet:
a cross-sectional view,
b preliminary design, and
c pole pitch adjustment design
[2]

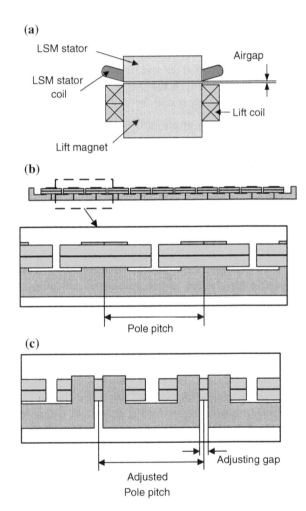

Fig. 5.20 Detent force versus position from initial arrangement [2]

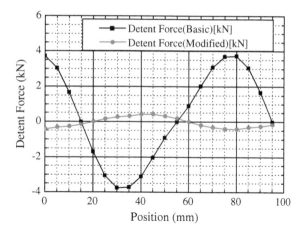

Fig. 5.21 Lift force versus position [2]

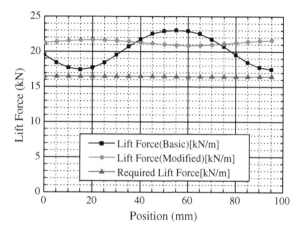

Fig. 5.22 Lift force versus current, with different airgaps [2]

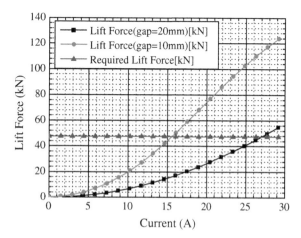

(a)

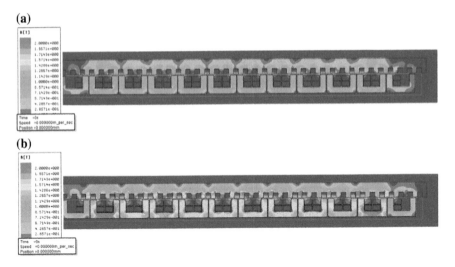

(b)

Fig. 5.23 Flux density distributions: **a** 17 A, 10 mm airgap, **b** 30 A, 20 mm airgap [2]

For design convenience, after an iterative design analysis of the 3-core magnet, the magnetic field analysis for the 12 poles full magnet module was performed. The flux distributions are obtained at two different currents and airgaps, i.e. nominal point (17 A, 10 mm airgap) and marginal point (30 A, 20 mm airgap) (Fig. 5.23). For the nominal point, it can be noted from Fig. 5.23 that the flux is equally distributed on each magnet thanks to the separation of magnets at a distance of 7 mm. This uniform flux distribution reduced the variations in thrust and levitation force while achieving the required lift force.

The levitation force-airgap characteristics of the magnet module are given in Fig. 5.24. The required lift force from one magnet module is 47 kN, and without

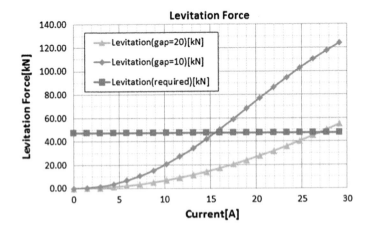

Fig. 5.24 Levitation force-current-airgap characteristics [2]

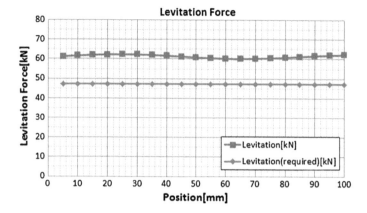

Fig. 5.25 Levitation force-position characteristics at 17 A, 10 mm airgap [2]

external disturbances, this can be achieved at around 17 A. Even at an airgap of 20 mm, the magnet module can provide the required force for levitation with 30 A.

Because of the uneven reaction surface, the levitation force varies with respect to the magnet's position relative to the stator, as shown in Fig. 5.25. This suggests that the varying lift forces are sufficiently above the required force. In addition, the levitation force ripple appears to not be clear, resulting in reduced levitation force disturbances.

Since the operation of LSM affects the levitation and thrust, the two forces need to be analyzed when LSM operates, i.e. when the currents flow in the LSM winding. The levitation ripple increases with the amount of current and needs to be considered in the levitation control loop, though the variations are small compared to when the current flows (Fig. 5.26).

On the other hand, because the flux of the lift magnet is used as a field system for LSM, the currents in the magnet coil influence the thrust. The variations in the thrust

Fig. 5.26 Levitation and thrust force-position with currents in LSM stator [2]

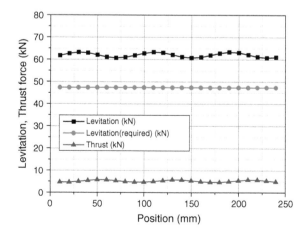

Fig. 5.27 Thrust
force-current angle-current
characteristics

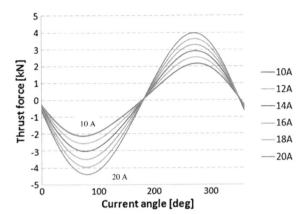

Fig. 5.28 Levitation
force-current angle-current
characteristics

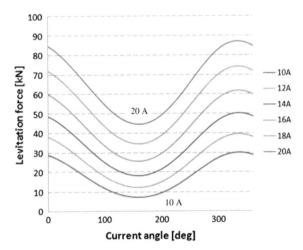

of LSM need to be assessed with the varying currents in the magnet coil. The results
of a simulation suggest that the thrust and levitation forces are almost proportional to
the increasing currents in the magnet coil. However, the current angle that gives
maximum thrust appears to be almost constant (Figs. 5.27 and 5.28).

- **U-core guidance magnet**: The lateral guidance force for the SUMA550 is
  provided via the U-core magnets with a flat reaction surface made of steel
  (Fig. 5.29). Allowing for the vehicle weight and speed, the design specifications
  of the magnet are listed in Table 5.4. To reduce the magnetic drag, the magnet
  center axis is aligned with the length of the vehicle, i.e. longitudinal direction.
  Considering weight, space and heat conduction, aluminum sheet coils were
  chosen.

Fig. 5.29 Configuration and dimensions of the guidance magnet for the SUMA550

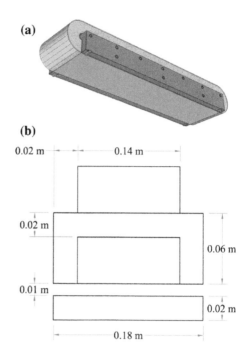

Table 5.4 Design specifications of the guidance magnet

Parameters	Values
Required guidance force (airgap = 20 mm)	7 kN/m
Guidance rail material	Steel
Guidance rail thickness	20 mm
Coil material	Aluminum
Turns (N)	120
Coil area	0.3 mm × 16 mm (sheet coil)
Nominal airgap	10 mm
Marginal airgap	20 mm
Number of magnets	16
Number of guidance magnets	16

Electromagnetic fields were evaluated in 2D using a commercial software package. It was found that the magnet has less leakage flux and was not saturated even when large currents established (Fig. 5.30). In addition to the guidance force, some guidance forces are usually offered by the fringe flux of the levitation magnet. For higher speeds, the guidance forces are sufficiently provided considering the curve radius and cant. The guidance forces are shown in Fig. 5.31, with increasing current at different airgaps.

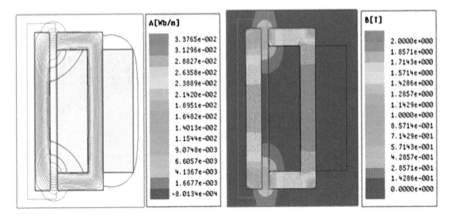

Fig. 5.30 Flux density distribution of the guidance magnet at 140 A, 10 mm airgap

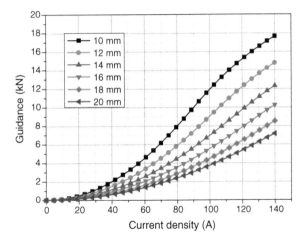

Fig. 5.31 Guidance force-current density characteristics with different airgaps

The effects of the thickness of the reaction rail on the guidance magnet on the flux distribution are analyzed, as shown in Fig. 5.32. The guidance forces were compared at airgaps of 30 mm and 20 mm. The differences in the airgap flux densities appear to be very slightly different. The guidance force is 6737 N for 30 mm thickness and 6652 N for 20 mm thickness, the difference being less than 2 %. For this reason, 20 mm was chosen as the thickness of the reaction rail for the guidance magnet.

Here, it may be relevant to compare the results from the 2D model with those from 3D. For the magnetic field analysis in 3D, the 2D model was extended to 3D

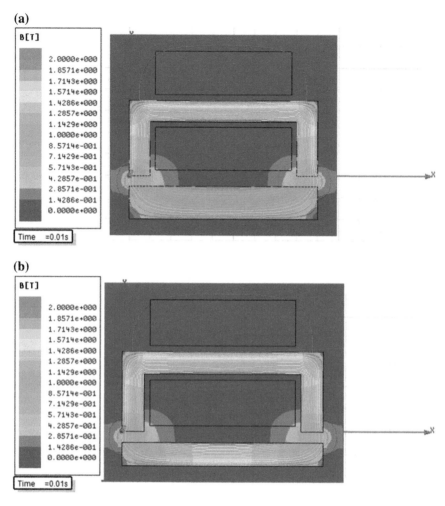

Fig. 5.32 Flux distribution: **a** thickness 30 mm, **b** thickness 20 mm

with the same boundary conditions. Figures 5.33 and 5.34 show the flux distribution and guidance forces, respectively. They suggest that the differences are very small—within 0.4 %—and thus the 2D model had sufficient accuracy to use.

Since the attraction force between the magnetized objects is directly related to the inductance in the magnetic circuit, the variations in the inductance need to be studied with the airgaps and currents in the circuit. The inductance variations are shown in Fig. 5.35, with different airgaps of 10 and 20 mm. In the case of a 10 mm airgap, the inductance is kept almost constant until 2.4 A/mm^2 but rapidly decreases after that current density. When the current density is 2.5 A/mm^2, the iron core appears to be severely saturated, as shown in Fig. 5.36.

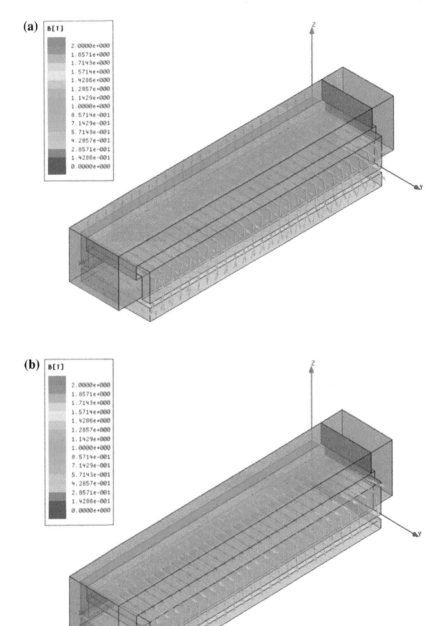

Fig. 5.33 Flux distribution in 3D: **a** thickness 30 mm, **b** thickness 20 mm

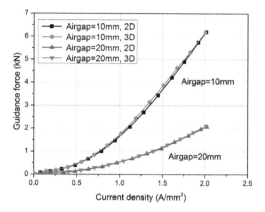

Fig. 5.34 Comparison of the guidance forces from 2D and 3D models

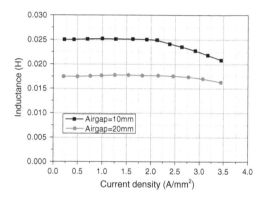

Fig. 5.35 Inductance-current density characteristics

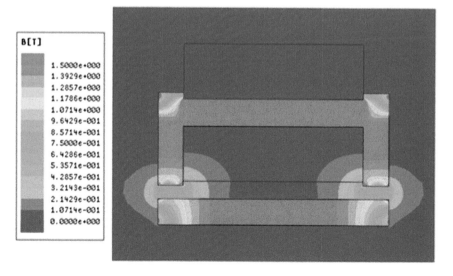

Fig. 5.36 Flux density distribution at airgap 10 mm, current density 2.5 A/m^2

5.4.4 Hybrid Magnets

One of the approaches to using permanent magnets to support vehicles is the use of the so-called hybrid magnet (or controlled permanent magnet), which is a combination of iron core and permanent magnet core, as shown in Fig. 5.37. The main advantage of this magnet is the considerable improvement in the lift/weight ratio and the reduction in the rated on-board power amplifier and supply. Thus, a brief analysis of the hybrid magnet is considered here, with a view to deriving the lift to power and lift to weight ratios. Using the dimensions and geometry in Fig. 5.37, the following expressions are obtained [1]. Note that the reluctance of the magnet core and the track were not considered.

Current in each coil:

$$I = k\bar{w}hJ$$

Airgap flux density:

$$B = \frac{\mu_0(I + H_m\bar{w})}{z}$$

Flux density in the permanent magnet core:

$$B_m = \frac{z}{d}B$$

Total lift force:

$$\bar{F} = \frac{B^2}{\mu_0}(pl)$$

$$\bar{F} = \frac{\mu_0 pl}{z^2}[k\bar{w}hJ + H_m\bar{w}]^2 \tag{5.23}$$

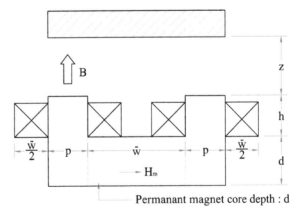

Fig. 5.37 Hybrid magnet geometry

Power dissipation in two magnet windings

$$\overline{P}_0 = 2I^2 r$$

where

$$r = \varrho \, \frac{mean \; length \; of \; each \; winding}{each \; winding \; area}$$

$$= \varrho \, \frac{\{2l + \pi(p + \bar{w})\}}{k\bar{w}h} \tag{5.24}$$

Therefore, winding copper loss

$$\overline{P}_0 = 2\varrho \, \frac{\{2l + \pi(p + \bar{w})\}}{k\bar{w}h} (k\bar{w}J)^2$$

$$= 2\varrho k\bar{w}h\{2l + \pi(p + \bar{w})\}J^2 \tag{5.25}$$

which is consistent with Eq. (5.22) with $2\bar{w} = w$. Total weight of the magnet

$$W_m = W_{pm} + W_i + W_w \tag{5.26}$$

where

$$W_{pm} = \text{weight of the permanent magnet block}$$
$$W_i = \text{weight of the magnet iron limbs}$$
$$W_w = \text{weight of the magnet windings}$$

Combining Eqs. (5.23)–(5.26), the lift force to magnet weight ratio

$$\overline{R}_{lw} = \frac{\overline{F}}{W_m} = \frac{\mu_0 p l (k\bar{w}hJ + H_m \bar{w})^2}{z^2 \left[W_{pm} + W_i + W_w \right]} \tag{5.27}$$

and the lift force to copper loss ratio

$$\overline{R}_{lp} = \frac{\overline{F}}{\overline{P}_0} = \frac{\mu_0 p l (k\bar{w}hJ + H_m \bar{w})^2}{2\varrho k\bar{w}h[2l + \pi(p + \bar{w})]J^2 z^2} \tag{5.28}$$

A comparison of Eqs. (5.21), (5.22), (5.27), and (5.28) confirms the expected improvements in $\overline{R}_{lw}$ and $\overline{R}_{lp}$ by combining the permanent magnet core with the iron core. Although a detailed analysis of the characteristics of the hybrid magnet is not included here, the main operating features are discussed based on earlier studies. While the weight reduction of the hybrid magnet is expected to be considerable, the benefit this provides needs to be weighed considering the significantly higher leakage flux in permanent magnets. The effects of external magnetic fields on the permanent magnet provide a limitation on their operations. Moreover, since allowable control power is very small compared to a normal electromagnet, the hybrid magnet may only be suitable where the airgap variation is very small, or levitated weight is relatively small. Currently, there are no maglev vehicles with this magnet in operation, though some are proposed. A maglev conveyor using the hybrid magnet has been built and introduced in Chap. 7.

5.4.5 Time Constant

As well as physical dimensions, the characteristics of the electromagnet are also considerably affected by the electrical properties of the core material and excitation winding. The electrical time constant (L/R) influences the dynamic responses to control input derived by power amplifier, and nonlinear electromagnetic field distribution becomes more important at higher frequency excitations. The former is directly related to resistance and inductance as well as the permeability of the core material. The latter is related to eddy currents in the magnet core and guideway. Earlier studies have found that there is a time lag between applied field and resulting airgap flux density, and it depends on the thickness (t_c), electrical resistivity and permeability of core material [1]. The relationship has been found to be

$$\Delta t = \frac{\pi \mu t_c^2}{3\varrho}$$

This implies that the time delay may be reduced by using laminated core, t_c being minimized. Since the number of coil turns of the excitation winding is directly related to m.m.f., the resistance and inductance of the coil is closely related to the airgap flux. From the above relations, it can be noted that the magnet's electrical time constant can be reduced by a smaller number of turns, a smaller pole face area, as well a large airgap, all of which reduce levitation force significantly. One of the methods for increasing m.m.f. is to use high input voltage to the magnet coil. This in turn increases the coil current, and thus increases m.m.f. with the identical coil turns. This method for reducing the effective time constant of the magnet winding is called voltage forcing. The allowable maximum voltage is related to the coil inductance (L) and maximum attainable dI/dt as

$$V_{forcing} = L \left[\frac{dI}{dt} \right]_{max}$$

All the expressions for the forces produced by electromagnets here are derived based on some simplifications and assumptions. For a more accurate dynamic analysis of electromagnets, particularly at higher frequency, nonlinear electromagnetic field distribution needs to be investigated. To do this, B–H curves for magnets and guideway may be necessary for numerical simulations based on a 3D model.

5.4.6   Eddy-Current

The expressions derived in the preceding chapters for the electromagnets are valid for stationary magnets. If a magnet moves relative to a ferromagnetic guideway, then the effects of the eddy currents induced by the moving magnet in the guideway on the dynamic behavior need to be considered to correct the force expressions with respect to varying relative speeds. The drag and repulsive forces are essentially generated by the eddy currents. The drag force requires additional propulsion forces, while the repulsion forces reduce the effective vertical attraction force. Hence, two forces caused by relative movements may lower the energy efficiency. For this reason, it is necessary to quantify the two forces. However, a brief review is introduced in this section rather than detailed derivations of these. Some suggestions and discussions are given with resulting expressions and numerical simulations. According to earlier studies, the ideal lift force (F_0) and actual lift force ($\widehat{F}_l$) corrected by eddy currents in the guideway of Fig. 5.38 are related through the following expression [1].

$$\widehat{F}_l(v) = F_0 \frac{8zl}{\mu_0 \sigma b p^2 v} \tag{5.29}$$

where

v: velocity
σ: constant ohmic conductivity
$p = 2\bar{p}$
$b = 2a$

Equation (5.29) gives the following observations.

- The lift force loss caused by eddy current effects could be reduced by using a narrower guideway ($l \gg b$). That is, the use of a narrow and long magnet reduces the loss of lift. In practice, several meter long magnets are placed along the length of vehicle.

Fig. 5.38 Magnet-guideway configuration for eddy-current analysis

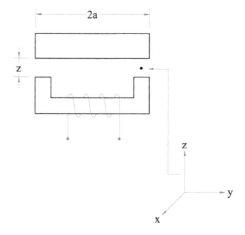

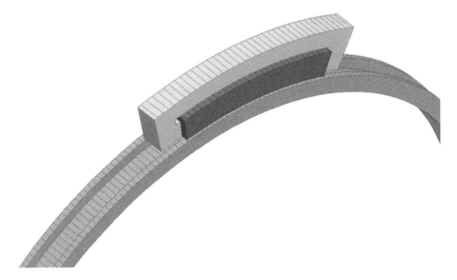

Fig. 5.39 A roller-rig model for eddy current effect analysis

- The ratio of z/l has a considerable influence on eddy current effects and thus a very small ration (i.e. $l = 100z$) is chosen for high speed vehicles. As can be expected, the loss of lift forces should be considered where a magnet moves at higher speeds.

The speed dependency of lift and drag forces are predicted through numerical simulations. The roller-rig model in Fig. 5.39 is employed with a narrow pole width U-core electromagnet. The pole width is 20 mm, length 750 mm, and airgap 10 mm. The variations of attraction and drag forces obtained from simulations are given in Fig. 5.40, which indicates that the force variations are less sensitive to speeds. This result suggests that a narrower pole width magnet should be used for high-speed vehicles.

Fig. 5.40 Repulsion and drag force variations with increasing speeds

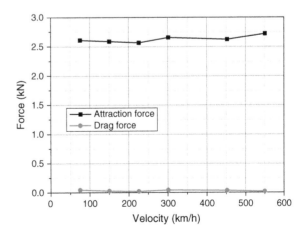

5.5 Suspension Control

The stabilization of the inherently unstable suspension in electromagnetic systems is the fundamental requirement of the levitation control systems. Though the generalized configurations of such suspension control loops were introduced in the preceding sections, there may be a variety of configurations and associated control schemes. Since the control law depends on a particular application, it would be adjusted considering the operating conditions and disturbances to it. This section introduces a basis for the stabilization of electromagnetic levitation systems. The system dynamic models are first presented, and then the implementation of the control laws that is applicable to the actual systems is demonstrated.

5.5.1 Dynamic Models

A more reasonable dynamic model is first needed for the design and analysis of a system, as well as the controller design and its performance assessment. The simplified dynamic models are developed and extended in this section [1]. It is assumed that the guideway is fixed to ground and remains unchanged, i.e. no deflection. Thus, the single magnet model in Sect. 5.2 is reused here (Fig. 5.41).

The flux linkage between the magnet and ferromagnetic guideway through airgap $z(t)$ is assumed to be ϕ_m. If the magnet geometry and winding are designed such that $\phi_T \cong \phi_m(\phi_l \cong 0)$, then the instantaneous magnet inductance is defined by

$$L(z, i) = \frac{N}{i}\phi_T = \frac{N}{i(t)}\frac{Ni(t)}{R_T} \tag{5.30}$$

Fig. 5.41 Single magnet-rigid guideway configuration

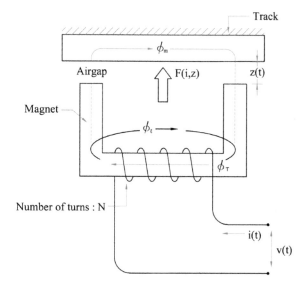

R_T in Eq. (5.30) is the total reluctance in the magnetic circuit. If the reluctance in the magnet core is compared to airgap (total length $= 2z(t)$), then the reluctance may be rewritten as

$$L(z) = \frac{\mu_0 N^2 A}{2z(t)} \tag{5.31}$$

In addition, the magnetic co-energy at any instant is $W_m(t) = (1/2)L(z)i(t)^2$. Then, the force of attraction between two objects is defined by

$$F(i,z) = \frac{dW_m(t)}{dz} = \frac{d}{dz}\left\{\frac{1}{2}L(z)i(t)^2\right\}$$
$$F(i,z) = \frac{B^2 A}{\mu_0} = \frac{\mu_0 N^2 A}{4}\left[\frac{i(t)}{z(t)}\right]^2 \tag{5.32}$$

With the total resistance R in the magnetic circuit, the magnet winding voltage $v(t)$ and the excitation current are related through the following expression.

$$\begin{aligned} v(t) &= Ri(t) + \frac{d}{dt}[L(z,i)i(t)] \\ &= Ri(t) + \frac{d}{dt}\left[\frac{\mu_0 N^2 A}{2z(t)}i(t)\right] \\ &= Ri(t) + \frac{\mu_0 N^2 A}{2}\frac{d}{dt}\left[\frac{i(t)}{z(t)}\right] \\ &= Ri(t) + \frac{\mu_0 N^2 A}{2z(t)}\frac{di(t)}{dt} - \frac{\mu_0 N^2 A i(t)}{2[z(t)]^2}\frac{dz(t)}{dt} \end{aligned} \tag{5.33}$$

Consequently, the required excitation current for levitation control may be obtained by $v(t)$. The equations of motion for the single-magnet are expressed as, with $f_d(t) =$ disturbance force input

$$\left.\begin{aligned} m\ddot{z}(t) &= -F(i,z) + f_d(t) + mg \\ mg &= F_0(i_0, z_0) = \frac{\mu_0 N^2 A}{4}\left[\frac{i_0}{z_0}\right]^2 \end{aligned}\right\} \tag{5.34}$$

where (i_0, z_0) represents the equilibrium (nominal) point. Though a precise dynamic analysis of this electromagnetic system could be achieved by solving Eqs. (5.32)–(5.34) numerically, a reasonable linear model may be useful for using well-established control theories. Thus, a linear model for the model in Fig. 5.41 needs to be obtained through numerical approximations. To derive the linear dynamic model, a linearized attraction force model may be used at the equilibrium point (i_0, z_0). The small perturbation linear equations may be defined with discounting second-order effects.

$$m\Delta\ddot{z}(t) = -\frac{\mu_0 N^2 A}{4}\left[\frac{i_0 + \Delta i(t)}{z_0 + \Delta z(t)}\right]^2 + f_d(t) + mg \tag{5.35}$$

$$\cong -\frac{\mu_0 N^2 A}{4}\left[\frac{i_0}{z_0}\right]^2\left[1 + 2\frac{\Delta i(t)}{i_0} - 2\frac{\Delta z(t)}{z_0}\right] + f_d(t) + mg$$

$$\cong -\frac{\mu_0 N^2 A i_0^2}{2z_0^2}\Delta i(t) + \frac{\mu_0 N^2 A i_0^2}{2z_0^3}\Delta z(t) + f_d(t)$$

$$\cong -k_i \Delta i(t) + k_z \Delta z(t) + f_d(t) \tag{5.36}$$

and

$$v_0 + \Delta v(t) = R[i_0 + \Delta i(t)] + \frac{\mu_0 N^2 A}{2}\frac{d}{dt}\left[\frac{i_0 + \Delta i(t)}{z_0 + \Delta z(t)}\right]$$

$$\cong R[i_0 + \Delta i(t)] + \frac{\mu_0 N^2 A}{2}\frac{d}{dt}\left\{\left[\frac{i_0}{z_0}\right]\left[1 + \frac{\Delta i(t)}{i_0} - \frac{\Delta z(t)}{z_0}\right]\right\}$$

$$\cong Ri_0 + R\Delta i(t) + \frac{\mu_0 N^2 A}{2z_0}\Delta i(t) - \frac{\mu_0 N^2 A i_0}{2z_0^2}\Delta \dot{z}(t)$$

$$\cong Ri_0 + R\Delta i(t) + L_0\Delta i(t) - k_i\Delta\dot{z}(t) \tag{5.37}$$

where, L_0 is the inductance of magnet winding at equilibrium point.

$$k_i = \left.\frac{\partial F(i,z)}{\partial i}\right|_{(i_0,z_0)}, \quad k_z = \left.\frac{\partial F(i,z)}{\partial z}\right|_{(i_0,z_0)}$$

A comparison of the coefficients in Eq. (5.36) and Eq. (5.37) suggests that

$$L_0 k_z = \frac{\mu_0 N^2 A}{2z_0} \times \frac{\mu_0 N^2 A i_0^2}{2z_0^3} = k_i^2 \tag{5.38}$$

Substituting k_i in Eq. (5.37), with $v_0 = Ri_0$, gives

$$\Delta\dot{i}(t) = \frac{k_z}{k_i}\Delta\dot{z}(t) - \frac{R}{L_0}\Delta i(t) + \frac{1}{L_0}\Delta v(t) \tag{5.39}$$

where, let β be

$$\beta = \frac{\mu_0 N^2 A}{2z_0}$$

By choosing $\Delta z(t)$, $\Delta\dot{z}(t)$ and $\Delta i(t)$ as state variables, the above equations may combined in the form of state-space representation as

$$
\begin{bmatrix} \Delta\dot{z}(t) \\ \Delta\ddot{z}(t) \\ \Delta i(t) \end{bmatrix} = \begin{bmatrix} 0 & 1 & 0 \\ \frac{k_z}{m} & 0 & -\frac{k_i}{m} \\ 0 & \frac{k_z}{k_i} & -\frac{R}{L_0} \end{bmatrix} \begin{bmatrix} \Delta z(t) \\ \Delta\dot{z}(t) \\ \Delta i(t) \end{bmatrix} + \begin{bmatrix} 0 & 0 \\ 0 & \frac{1}{m} \\ \frac{1}{L_0} & 0 \end{bmatrix} \begin{bmatrix} \Delta v(t) \\ f_d(t) \end{bmatrix} \qquad (5.40)
$$

Since $\Delta z(t)$ is the airgap between the magnet and the reaction surface, the above equations describe the dynamics of the suspended (or levitated) mass. That is, the dynamic equation represents the motion of the mass relative to the fixed guideway. This linear model may be suitable enough for the controller design and analysis if the guideway is fixed and its deflection is negligible. In practice, this simplified linear dynamic model may be the basis for stabilization through control loops. However, if the position varies or the magnet moves over an uneven guideway surface, the position of magnet should be represented with respect to an absolute datum line, leading to the modification of the dynamic equations based on absolute position, not relative position. When there is a moving magnet and an uneven guideway surface, as seen in Fig. 5.42, an absolute datum is introduced. Variations in guideway height occur due to the following reasons.

- Static deflection of the elevated guideway beams due to the moving mass
- Guideway irregularities due to uneven joints and rough surface finishing
- Deflection of the guideway due to dynamic interaction between the modes of the beam and the vehicle suspension system

Since the waveforms of those causes have relatively different amplitudes and frequencies, the influences of those causes on system dynamics are also different. The static and dynamic deflection of the guideway has the most significant effects

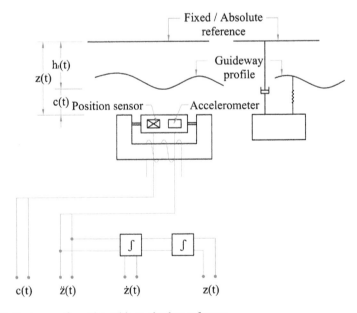

Fig. 5.42 System configuration with an absolute reference

on suspension stability because of their large amplitude and lower frequency than irregularities. As such, these two deflections are treated in the levitation control part and the guideway surface roughness is considered in the ride control part. The dynamic interaction will be introduced in a dedicated section later because it significantly affects both suspension stability and guideway construction costs. Since the vertical acceleration related to ride comfort and the airgap related to stability are the basic design parameters of electromagnetic systems, the vertical acceleration ($\ddot{z}(t)$) measured with respect to the absolute datum line and the airgap from the guideway height ($h_t(t)$) to the magnet position $c(t)$ need to be combined in the equations of motion for the system. If $\Delta h_t(t)$ represents the deviation from the nominal height h_0 and $\Delta c(t)$ the deviation from the nominal airgap c_0, then Eq. (5.36) may be redefined as

$$m\Delta\ddot{z}(t) = -k_i\Delta i(t) + k_z\Delta z(t) + f_d(t) \tag{5.41}$$

where the magnet current is related to input voltage and relative motion

$$m\Delta i(t) = \frac{k_z}{k_i}\Delta\dot{c}(t) - \frac{R}{L_0}\Delta i(t) + \frac{1}{L}\Delta z(t) \tag{5.42}$$

By choosing $\Delta c(t), \Delta\dot{z}(t)$ and $\Delta i(t)$ as the new state variables and $\Delta h_t(t)$ and $\Delta v(t)$ as forcing functions, the above equations may be combined in the state space form

$$\begin{bmatrix} \Delta\dot{c}(t) \\ \Delta\ddot{z}(t) \\ \Delta\dot{i}(t) \end{bmatrix} = \begin{bmatrix} 0 & 1 & 0 \\ \frac{k_z}{m} & 0 & -\frac{k_i}{m} \\ 0 & \frac{k_z}{k_i} & -\frac{R}{L_0} \end{bmatrix} \begin{bmatrix} \Delta c(t) \\ \Delta\dot{z}(t) \\ \Delta i(t) \end{bmatrix} + \begin{bmatrix} 0 & 0 \\ 0 & \frac{1}{m} \\ \frac{1}{L_0} & 0 \end{bmatrix} \begin{bmatrix} \Delta v(t) \\ f_d(t) \end{bmatrix} + \begin{bmatrix} -1 \\ 0 \\ -\frac{k_z}{k_i} \end{bmatrix} \Delta\dot{h}_t(t)$$

$$\tag{5.43}$$

There are two position signals, i.e. $\Delta c(t)$ and $\Delta z(t)$. From a levitation stability viewpoint, $\Delta c(t)$, a relative position, must be maintained within a specified range. And $\Delta z(t)$, an absolute position, is desirably small for the sake of ride comfort. As such, there will be a compromise between them with frequency, because ride comfort depends on frequency. The desirable property of $z_0 + \Delta z(t)$ is that it follows the guideway profile at lower frequencies, keeping $\Delta c(t)$ within the allowable range, while at higher frequencies it becomes less sensitive to guideway height variations, absolute vertical acceleration being small. This latter requirement suggests that at higher frequencies, the suspended mass must maintain an absolute height $z_0 = c_0 + h_0$ with respect to the fixed datum.

5.5.2 Open-Loop System

Before considering the closed-loop control for levitation, the open-loop system needs to be studied. The linear equations of motion derived in the preceding section are used for open-loop system analysis.

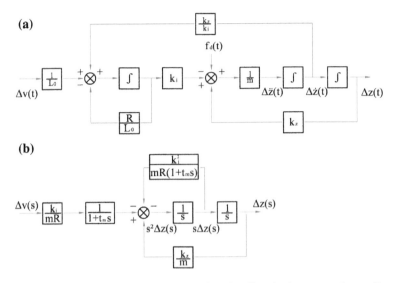

Fig. 5.43 Block diagram of the open-loop system based on linearized state equations. **a** Based on linearized state equations. **b** Modified transfer function

$$
\begin{bmatrix} \Delta \dot{z}(t) \\ \Delta \ddot{z}(t) \\ \Delta i(t) \end{bmatrix} = \begin{bmatrix} 0 & 1 & 0 \\ \frac{k_z}{m} & 0 & -\frac{k_i}{m} \\ 0 & \frac{k_z}{k_i} & -\frac{R}{L_0} \end{bmatrix} \begin{bmatrix} \Delta z(t) \\ \Delta \dot{z}(t) \\ \Delta i(t) \end{bmatrix} + \begin{bmatrix} 0 & 0 \\ 0 & \frac{1}{m} \\ \frac{1}{L_0} & 0 \end{bmatrix} \begin{bmatrix} \Delta v(t) \\ f_d(t) \end{bmatrix} \tag{5.44}
$$

The block diagram for Eq. (5.44) may be made as in Fig. 5.43 [1].

The transfer function for this open-loop system is expressed as, with $f_d(t) = 0$

$$
\begin{aligned}
\Delta Z(s) &= \begin{bmatrix} 1 & 0 & 0 \end{bmatrix} \frac{Adj(sI - A)}{det(sI - A)} \begin{bmatrix} 0 \\ 0 \\ \frac{1}{L_0} \end{bmatrix} \Delta V(s) \\[2mm]
&= -\frac{(k_i/mL_0)}{s^3 + \frac{R}{L_0}s^2 - \frac{k_z}{m}\frac{R}{L_0}} \Delta V(s) \\[2mm]
&= -\left[\frac{k_i/mR}{\frac{L_0}{R}s^3 + s^2 - \frac{k_z}{m}} \right] \Delta V(s) \\[2mm]
&= -\left[\frac{(k_i/mR)}{\left(1 + \frac{L_0}{R}s\right)\left\{ s^2 + \frac{k_i^2}{mR\left(1 + \frac{L_0}{R}s\right)}s - \frac{k_z}{m} \right\}} \right] \Delta V(s) \tag{5.45}
\end{aligned}
$$

If the power amplifier has a wide enough bandwidth $((t_m = L_0/R) \ll 1)$, then the approximation of the open-loop characteristic equation may be derived as

$$(1 + t_m s)\left(s^2 + \frac{k_i^2}{mR}s - \frac{k_z}{m}\right) = 0 \tag{5.46}$$

This equation gives open-loop poles

$$s_1 = -\frac{1}{t_m}$$

$$s_{2,3} = \frac{1}{2}\left[-\frac{k_i^2}{mR} \pm \sqrt{\left(\frac{k_i^4}{m^2R^2} + 4\frac{k_z}{m}\right)}\right]$$

$$= \frac{1}{2}\left[-\frac{L_0 k_z}{mR} \pm \sqrt{\left(\frac{L_0^2 k_z^2}{m^2R^2} + 4\frac{k_z}{m}\right)}\right]$$

$$= -\frac{t_m k_z}{2m} \pm \sqrt{\left[\left(\frac{t_m k_z}{2m}\right)^2 + \frac{k_z}{m}\right]}$$

These open-loop poles indicate that the mechanical poles are real and closely related to the electrical pole of the magnet amplifier. The effect of a non-zero t_m is to shift the open-loop poles towards the right of the real axis by a distance that is greater than $(t_m k_z/2m)$. Consequently, it can be noted that t_m has profound effects on the dynamic characteristics in electromagnetic systems, i.e. control performances. Since the constants k_i, k_z depend on $\partial F(\cdot)/\partial i$, $\partial F(\cdot)/\partial z$ at nominal point (i_0, z_0) and m depends on the nominal weight of the levitated object, the dynamics are highly dependent on the selection of the nominal equilibrium point. The reason for this is that k_i, k_z vary differently with the nominal point due to the non-linear nature of attraction force. This also implies that the operating point must be significantly closer to the nominal point in the electromagnetic system. If (i_0, z_0) is on the non-linear part of magnet force-current-distance characteristics, maximum permissible perturbation of the airgap is severely limited in order to maintain linearity. This requirement may be met by choosing (i_0, z_0) on the linear part of magnet characteristics, and by using a position control system with considerably high loop gain to make the steady-state airgap errors very small.

5.5.3 Closed-Loop System

In the first analysis of suspension stability, the guideway may be assumed to be a rigid body. Based on this assumption, the equations of relative motion Eq. (5.44) and its transfer function Eq. (5.45) are used as a basis for the design of closed-loop control. The transfer function Eq. (5.45) may be rearranged into a typical form.

$$\Delta Z(s) = -\frac{(k_i/mL_0)}{s^3 + \frac{R}{L_0}s^2 - \frac{k_z R}{m L_0}} \Delta V(s) \qquad (5.47)$$

The above transfer function suggests that for a given mass m, k_i controls the input-output gain of the open-loop system, while k_z controls its poles. The locations of the open-loop poles are given by

$$s^3 + \frac{R}{L_0}s^2 - \frac{k_z R}{m L_0} \equiv (s+\gamma)(s^2 + \alpha s - \gamma) = 0 \qquad (5.48)$$

where the coefficients α and γ are related through the non-linear relationship as

$$\alpha + \gamma = \frac{R}{L_0} \quad \text{and} \quad \alpha\gamma^2 = \frac{k_z R}{m L_0} \rightarrow \alpha\gamma^2 = \frac{k_z}{m}(\alpha + \gamma) \qquad (5.49)$$

Since $k_z \propto i_0^2/z_0^3$, the deviation from the chosen operating point (i_0, z_0) will change k_z significantly, and in turn move open-loop poles. Thus, the electromagnet system should be designed such that the magnets operate satisfactorily at very close to the nominal equilibrium point. The property of the shift of the nominal point is well illustrated in Fig. 5.44. The two poles are located in a stable region on the left half plane, and one pole is placed in an unstable region on the right half plane. As k_z increases, both the convergence and frequency of the two poles on the left half

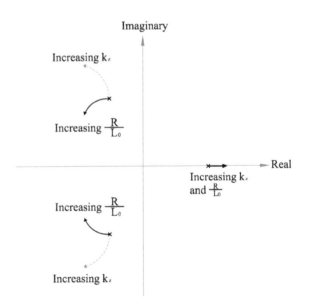

Fig. 5.44 Movement of open-loop poles owing to changing k_z and R/L_0

plane are increased, while the divergence of the one pole on the right half plane increased. If R/L_0 becomes larger, meaning that the time constant is smaller, the frequencies of two poles are decreased, with increasing convergence. And the divergence of one pole becomes rapid. If the m varies, the movement of the open-loop poles would be much more involved. Due to these considerable effects of parameters in electromagnet systems, it must be ensured that the lift magnet operates very close to the nominal position. The open-loop poles in Fig. 5.44 suggest that at least one zero is needed if the system is to be stabilized using the classical compensation techniques. To stabilize this inherently unstable system, an appropriate closed-loop needs to be incorporated. Of course, there is a wide range of close-loop schemes that could be applicable to this system, of which two representative methods, flux feedback and state feedback, are briefly presented in the following sections.

5.5.4 Flux Feedback

As mentioned in the previous section, due to the significant effect of k_z on the closed-loop poles, the controlled electromagnet system may not have adequate stability if the operating point is changed from the nominal point. Relating the compensator zero (z_c) and the design parameters to the closed-loop stability analysis would be a considerably involved process. Thus, before achieving the desired closed-loop characteristics, it may first be desirable to make the system dynamics less sensitive to variations in operating point. As a way of accomplishing this, an airgap flux feedback scheme may be used to reduce the effects of k_z, k_i on suspension dynamics. The concept of the airgap flux feedback may be obtained by slightly modifying the block diagram in Fig. 5.45 [1].

Since the flux linkage between the magnet and the track is linearly proportional to the magnet excitation current and inversely proportional to the airgap, the small perturbation model around the nominal point ϕ_0 may be expressed by

$$\Delta\phi(t) = k_{fi}\Delta i(t) - k_{fz}\Delta z(t)$$

Consequently, if a proportion of flux is fed into the power amplifier, as shown in Fig. 5.45, Eq. (5.39) for the small perturbation magnet excitation current becomes

$$\begin{aligned}
\Delta\dot{i}(t) &= \frac{k_z}{k_i}\Delta\dot{z}(t) - \frac{R}{L_0}\Delta i(t) - \frac{k_\phi}{L_0}\left[k_{fi}\Delta i(t) - k_{fz}\Delta z(t)\right] + \frac{1}{L_0}\Delta v(t) \\
&= \frac{k_z}{k_i}\Delta\dot{z}(t) + \frac{k_\phi}{L_0}k_{fz}\Delta z(t) - \left[\frac{k_\phi}{L_0}k_{fi} + \frac{R}{L_0}\right]\Delta i(t) + \frac{1}{L_0}\Delta v(t)
\end{aligned} \tag{5.50}$$

With the choice of state variables as in Eq. (5.40), Eqs. (5.36) and (5.50) give

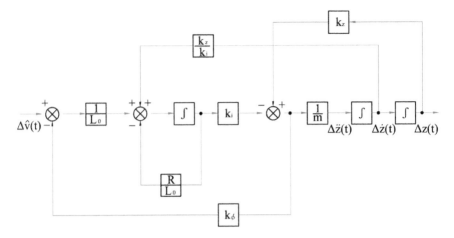

Fig. 5.45 Open-loop system with flux feedback

$$\begin{bmatrix} \Delta\dot{z}(t) \\ \Delta\ddot{z}(t) \\ \Delta i(t) \end{bmatrix} = \begin{bmatrix} 0 & 1 & 0 \\ \frac{k_z}{m} & 0 & -\frac{k_i}{m} \\ \frac{k_\phi}{L_0}k_{fz} & \frac{k_z}{k_i} & -\frac{k_\phi}{L_0}k_{fi} - \frac{R}{L_0} \end{bmatrix} \begin{bmatrix} \Delta z(t) \\ \Delta\dot{z}(t) \\ \Delta i(t) \end{bmatrix} + \begin{bmatrix} 0 & 0 \\ 0 & \frac{1}{m} \\ \frac{1}{L_0} & 0 \end{bmatrix} \begin{bmatrix} \Delta v(t) \\ f_d(t) \end{bmatrix} \quad (5.51)$$

Though the input–output gain remains unchanged, the characteristics equation is modified.

$$det\{sI - A_f\} = s^3 + \left(\frac{R}{L_0} + \frac{k_\phi}{L_0}k_{fi}\right)s^2 + \frac{k_\phi}{mL_0}\left(k_ik_{fz} - k_zk_{fi}\right) - \frac{k_z}{m}\frac{R}{L_0} \quad (5.52)$$

Thus if the gain of flux feedback loop is adjusted such that, to eliminate $\frac{k_z}{m}\frac{R}{L_0}$,

$$k_\phi\left(k_ik_{fz} - k_zk_{fi}\right) = k_z \quad (5.53)$$

then the input–output transfer function reduces to (from Eqs. (5.52) and (5.53) with $f_d(t) = 0$)

$$\Delta Z(s) = -\frac{k_i/mL_0}{s^2\left(s + \frac{R}{L_0} + k_\phi k_{fi}\right)}\Delta V(s) \quad (5.54)$$

This transfer function indicates that the flux feedback makes the system conditionally stable. This result has considerable practical significance, as the value of gain k_ϕ can be fairly easily derived through experiments. This conditionally stable electromagnetic system tends to be attracted to the guideway with very soft stiffness. From an analytical viewpoint, the flux loop makes the open-loop poles independent

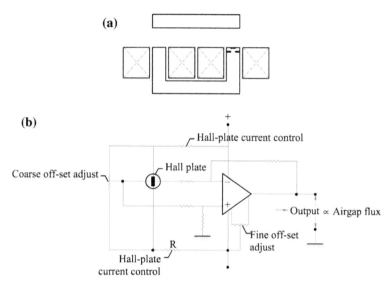

Fig. 5.46 Location of Hall plates for flux measurement

of k_z and as a result the compensator configuration becomes less sensitive to the nominal operating airgap. However, the sensitivity of the loop gain to k_i remains, i.e. still sensitive to the nominal current. To implement this control method, the flux needs to be measured by placing Hall plates as shown in Fig. 5.46 [1].

5.5.5 States Feedback

If the electromagnetic levitation system is to be viable, then it is desirable for the closed-loop control system to be capable of considering the deflection and irregularity of the guideway, as well as the dynamic interaction between vehicle and guideway. A well-known states feedback control scheme is one of the control laws that is relatively general and includes as many states as possible. This scheme may be used as a basis for a particular application. The dynamic equations with respect to absolute reference are reused here to introduce the states feedback control scheme.

$$\begin{bmatrix} \Delta\dot{c}(t) \\ \Delta\ddot{z}(t) \\ \Delta i(t) \end{bmatrix} = \begin{bmatrix} 0 & 1 & 0 \\ \frac{k_z}{m} & 0 & -\frac{k_i}{m} \\ 0 & \frac{k_z}{k_i} & -\frac{R}{L_0} \end{bmatrix} \begin{bmatrix} \Delta c(t) \\ \Delta\dot{z}(t) \\ \Delta i(t) \end{bmatrix} + \begin{bmatrix} 0 & 0 \\ 0 & \frac{1}{m} \\ \frac{1}{L_0} & 0 \end{bmatrix} \begin{bmatrix} \Delta v(t) \\ f_d(t) \end{bmatrix} + \begin{bmatrix} -1 \\ 0 \\ -\frac{k_z}{k_i} \end{bmatrix} \Delta\dot{h}_t(t)$$

$$(5.55)$$

A feedback control law using all the states in Eq. (5.55) for stabilization may be defined by

$$\Delta v(t) = k_{zpp}\Delta\ddot{z}(t) + k_{zp}\Delta\dot{z}(t) + k_z\Delta z(t) + k_{gp}\Delta\dot{c}(t) + k_g\Delta c(t) \qquad (5.56)$$

The control voltage is derived by multiplying 5 states by corresponding 5 gains, as in Eq. (5.56). To study the feedback of 5 states, substituting Eq. (5.56) into Eq. (5.55) gives (with $\Delta c(t) = \Delta z(t) + \Delta h_t(t)$, $\Delta\dot{c}(t) = \Delta\dot{z}(t) - \dot{h}_t(t)$, and $\Delta h_t(t) = \int \Delta\dot{h}_t(t)dt$)

$$
\begin{bmatrix} \Delta\dot{c}(t) \\ \Delta\ddot{z}(t) \\ \Delta i(t) \end{bmatrix} = \begin{bmatrix} 0 & 1 & 0 \\ \frac{k_c}{m} & 0 & -\frac{k_i}{m} \\ \left(\frac{k_p}{L_0} + \frac{k_c k_a}{mL_0}\right) & \left(\frac{k_v}{L_0} + \frac{k_c}{k_i}\right) & -\left(\frac{R}{L_0} + \frac{k_i k_a}{mL_0}\right) \end{bmatrix} \begin{bmatrix} \Delta c(t) \\ \Delta z(t) \\ \Delta i(t) \end{bmatrix}
$$
$$
+ \begin{bmatrix} -1 \\ 0 \\ -\frac{k_c}{k_i} - \frac{k_{gp}}{L_0} + \frac{k_z}{L_0}\int dt \end{bmatrix} \Delta\dot{h}_t(t) \qquad (5.57)
$$

where $k_a = k_{zpp}$, $k_v = k_{zp} + k_{gp}$, $k_p = k_z + k_g$. The characteristics equation for Eq. (5.57) is as follows

$$p(s) = s^3 + \left(\frac{R}{L_0} + \frac{k_i k_a}{mL_0}\right)s^2 + \frac{k_i k_v}{mL_0}s + \frac{k_i k_p - k_c R}{mL_0} \qquad (5.58)$$

For the stabilization of the system, the 5 gains in Eq. (5.56) may be chosen such that $k_p > k_c R / k_i$. If the electrical time constant is assumed to be much smaller than mechanical time constant, the characteristics equation may be approximated to a second-order system with the following natural frequency and damping ratio.

$$\widehat{w_n^2} = \frac{k_i k_p - k_c R}{k_c k_a + mR}, \quad 2\hat{\xi}\widehat{w_n} = \frac{k_i k_v}{k_i k_a + mR} \qquad (5.59)$$

k_p controls the response speed and k_v damping ratio. k_a influences both the response speed and damping ratio, but its effect is small. k_z and k_g of k_p have identical effects on the natural frequency; similarly, k_{zp} and k_{gp} of k_v affect the damping ratio to an identical extent. The control gains k_p, k_v and k_a may be derived by comparing the characteristics equation in Eq. (5.58) with the typical 3rd order polynomial $p(s) = (s + p_1)(s^2 + 2\xi w_n s + w_n^2)$. To study the effectiveness of the control gains, the transfer functions for the airgap $\Delta c(t)$ and absolute position $\Delta z(t)$ in terms of the guideway height $\Delta h_t(t)$ need to be investigated. The resulting transfer functions are

$$\frac{\Delta C(s)}{\Delta H_t(s)} = \frac{s^3 + \left(\frac{R}{L_0} + \frac{k_i k_a}{m L_0}\right) s^2 + \frac{k_i (k_v - k_{gp})}{m L_0} s + \frac{k_i k_z}{m L_0}}{p(s)} \tag{5.60}$$

$$\frac{\Delta Z(s)}{\Delta H_t(s)} = \frac{\frac{k_i k_{gp}}{m L_0} s + \frac{k_i (k_p - k_z) - k_c R}{m L_0}}{p(s)} \tag{5.61}$$

$\Delta C(s)/\Delta H_t(s)$ has the form of the transfer function for the high-pass filter and $\Delta Z(s)/\Delta H_t(s)$ the low-pass filter, respectively. And $\Delta C(s)/\Delta H_t(s) - \Delta Z(s)/\Delta H_t(s) = 1$ implies that two transfer functions have a complementary relationship. When k_z and k_{gp} are varied while k_p and k_v remain unchanged such that the characteristic equation of the overall control system is maintained, only the zeros in the above transfer functions, Eqs. (5.60) and (5.61), are changed. The frequency responses are shown in Figs. 5.47 and 5.48 with varying k_{gp}. Here, k_p, k_v and k_a were chosen, with $k_z = 0$, $p_1 = 2 \times \pi \times 10$, $\omega_n = 2 \times \pi \times 10$ rad/s, and $\xi = 0.707$.

If k_{gp} is increased, in the cut-off range the amplitudes of airgap deviation responses are relatively small, and the slope becomes steep as the cut-off frequency is approached (Fig. 5.47). This implies that at lower frequencies, the deviations in airgaps in terms of the guideway height variations may be reduced by increasing the control gain for the airgap velocity k_{gp}, resulting in as constant an airgap as possible. On the other hand, a larger k_{gp} gives a larger deviation in absolute position, reducing ride comfort (Fig. 5.48). This can be noted through a comparison of the absolute position responses for the absence and presence of the control gain k_{gp}.

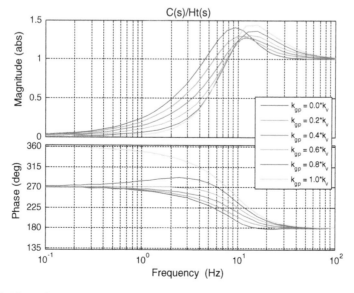

Fig. 5.47 Airgap frequency response of the system, with variations in gains for velocity

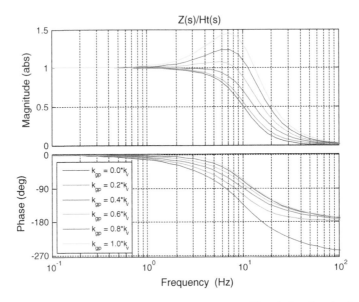

Fig. 5.48 Absolute position frequency response of the system, with variations in gains for velocity

$$\frac{\Delta Z(s)|_{k_{gp}\neq 0}}{\Delta Z(s)|_{k_{gp}=0}} = \frac{\frac{k_i k_{gp}}{mL_0}s + \frac{k_i\left(k_p-k_z\right)-k_c R}{mL_0}}{\frac{k_i\left(k_p-k_z\right)-k_c R}{mL_0}} \geq 1$$

This comparison shows that the absolute position deviation increases with k_{gp}. As can be seen in Fig. 5.48, when k_{gp} is used, the k_{gp} of 0–0.4 times k_v may be favorable for a lower change in the frequency responses. In this design, due to large deviations in the absolute position, a way to suppress the oscillations, e.g. through secondary suspension, may be introduced in order to provide an acceptable ride comfort. Next, the frequency responses of the deviations in airgap and absolute position in terms of the guideway height deviation are given in Figs. 5.49 and 5.50, with $k_{gp} = 0$, identical k_p, k_v,k_a in Figs. 5.47 and 5.48. In the presence of k_z, the changes in amplitude are highlighted at low frequencies. That is, the effects of the guideway height deviations, both on the airgap and absolute position, appeared with some proportions depending on k_z. This may eventually lead to steady-state errors in position and airgap, and thus it would be not suitable to use k_z, except for adjusting the absolute position.

From a practical viewpoint, a two-stage design procedure may be adopted for the implementation of the state feedback scheme. The first stage is concerned with the measurement of state variables. Since the direct measurement of all the state variables may pose practical difficulties, a dynamic filter (or state observer) may be employed to derive all the states only with least states measured. A method for deriving the 3 remaining states, in Eq. (5.56), which required that only airgap and acceleration be measured, may be successfully used in maglev vehicles. The details

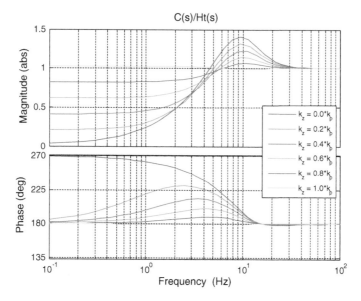

Fig. 5.49 Airgap frequency response of the system, with variations in gain for position

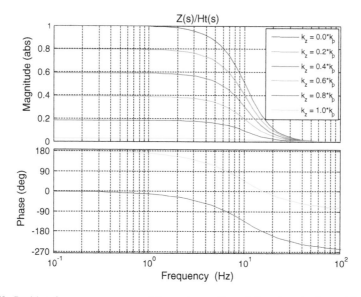

Fig. 5.50 Position frequency response of the system, with variations in gain for position

of the dynamic filter will be introduced below. In the second stage, the 5 control gains are adjusted considering all of the performance and operational requirements, which are dependent on a particular application. Some skills available during the second stage may be summarized as follows.

- Because one of the requirements of a maglev suspension is to follow the guideway profile while achieving acceptable ride comfort, a scheme may be adopted which follows the large amplitude deflections of the guideway at lower frequencies but ignores small amplitude surface roughness at higher frequencies. To apply this scheme, at lower frequencies, $\Delta c(t)$ and $\dot{\Delta c}(t)$ are highly weighted, while $\Delta z(t), \Delta \dot{z}(t)$ and $\Delta \ddot{z}(t)$ are highly weighted at higher frequencies. Weighting the absolute position $\Delta z(t)$ reduces the deviation of the absolute vertical position of the moving object (vehicle), resulting in an improvement in vertical acceleration levels (ride comfort). From a ride quality perspective, since the human body is sensitive to vertical acceleration at around 1 Hz, the choice of 1–2 Hz as a boundary between low and high frequency, i.e. cut-off frequencies of high- and low-pass filters, may be an appropriate selection.
- The control gains for positions determine the stiffness of a magnetic suspension.
- The control gains for relative and absolute velocities determine the damping, both producing identical portions of damping.
- If the relative velocity is weighted, then the phase lag between the airgap and the guideway height would be increased with increasing frequency.
- When the absolute velocity and airgap are weighted simultaneously, the desired frequency responses of the suspension system may be achieved.
- The use of acceleration is related to the effect of using flux feedback. The flux feedback equals the use of force of attraction.
- Airgap deviation integral effects may be added to the state feedback loop, Eq. (5.56), to reduce the steady-state airgap error regardless of the change in payload.
- Second or third time derivatives of the airgap also may be added to the state feedback loop, Eq. (5.56), to suppress the structural vibrations of the guideway structure.

5.5.6 Application of States Feedback

There may be various ways to derive the required states from the least measured states when making use of the state feedback scheme. As a proven application, the 5 states feedback scheme is applied to the experimental vehicle SUMA550 in the following sections [2]. As was described earlier, the first stage is to derive the 5 states in Eq. (5.56). The dynamic filter used here may have a state space representation as

$$\dot{\mathbf{x}} = \mathbf{A}\mathbf{x} + \mathbf{B}\mathbf{u}$$
$$\mathbf{y} = \mathbf{C}\mathbf{x} + \mathbf{D}\mathbf{u}$$

(5.62)

where,

$$u(t) = [Acc(t) \quad Gap(t)]^{\mathrm{T}}$$
$$y(t) = [\Delta\ddot{z}(t) \quad \Delta\dot{z}(t) \quad \Delta z(t) \quad \Delta\dot{c}(t) \quad \Delta c(t)]^{\mathrm{T}}$$

$Acc(t)$: measured acceleration

$Gap(t)$: measured airgap from nominal point

Matrices A, B, C, D in Eq. (5.62) are chosen based on the following skills.

- For obtaining $\Delta\ddot{z}(t)$, at low frequency the second time derivatives of the measured airgap $Gap(t)$ are used as $\Delta\ddot{z}(t)$ through a low-pass filter, and at high frequency the measured acceleration $Acc(t)$ is used as $\Delta\ddot{z}(t)$ with a high-pass filter. Then they are summed over the entire frequency range. The break frequency and damping ratio in the two filters are the key parameters in the two filters. Naturally, the selected parameters are dependent on the particular systems under consideration. In passenger vehicles, a break frequency of 1 Hz may be selected to remove DC components in the acceleration from an accelerometer. Instead, the second time derivative of the measured airgap is used.
- $\Delta\dot{z}(t)$ and $\Delta z(t)$ are derived in the same manner as $\Delta\ddot{z}(t)$.
- $\Delta\dot{c}(t)$ may be obtained by taking the time derivative of $Gap(t)$ filtered with a low-pass filter. The break frequency is selected as 100 Hz in passenger vehicles.
- $\Delta c(t)$ may be derived in the same manner as $\Delta\dot{c}(t)$.

This scheme is shown as a block diagram in Fig. 5.51, which is represented in the matrix form of

$$A = \begin{bmatrix} 0 & \frac{1}{T_3} & 0 & -\frac{1}{T_3} & 0 \\ -\frac{1}{T_1} & -\frac{V_1}{T_1} & 0 & \frac{V_1}{T_1} & 0 \\ 0 & \frac{1}{T_3} & -\frac{V_2}{T_2} & 0 & \frac{V_2}{T_2} \\ 0 & 0 & 0 & -\frac{V_3}{T_4} & -\frac{1}{T_4} \\ 0 & 0 & 0 & \frac{1}{T_5} & 0 \end{bmatrix} \quad B = \begin{bmatrix} 0 & 0 \\ \frac{1}{T_1} & 0 \\ 0 & 0 \\ 0 & \frac{1}{T_4} \\ 0 & 0 \end{bmatrix}$$

$$C = \begin{bmatrix} -1 & -V_1 & 0 & V_1 & 0 \\ 0 & 1 & 0 & 0 & 0 \\ 0 & 0 & 1 & 0 & 0 \\ 0 & 0 & 0 & 1 & 0 \\ 0 & 0 & 0 & 0 & 1 \end{bmatrix} \quad D = \begin{bmatrix} 1 & 0 \\ 0 & 0 \\ 0 & 0 \\ 0 & 0 \\ 0 & 0 \end{bmatrix}$$

where $T_i (i = 1, 2, 3, 4, 5)$ represents time constants which determine bandwidths for the filters, and $V_i (i = 1, 2, 3)$ give the damping ratios in the filters.

The 5 states feedback control scheme described earlier is applied to the experimental vehicle SUMA550. A magnet module consisting of 3 cores is studied to

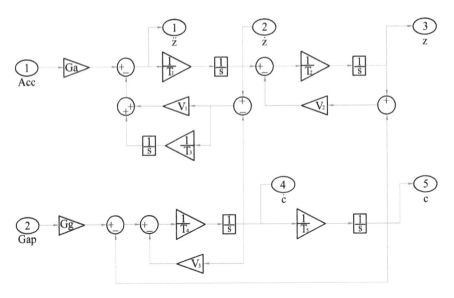

Fig. 5.51 Block diagram of the dynamic filter for obtaining 5 states

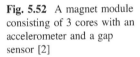

Fig. 5.52 A magnet module consisting of 3 cores with an accelerometer and a gap sensor [2]

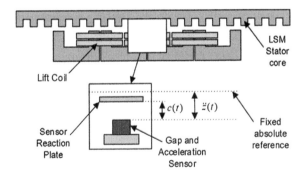

derive the control gains before extending to the full vehicle. One accelerometer and one airgap sensor are used to implement the dynamic filter in Fig. 5.52.

Table 5.5 gives the parameters and control gains chosen and the properties of the suspension control system. Using the values in Table 5.3, the transfer functions of the absolute acceleration, velocity and position to the measured acceleration are analyzed by plotting Bode diagrams. In addition, the transfer functions of the airgap and its velocity to the measured airgap are also plotted. From the Bode diagrams in Fig. 5.53, it can be noted that in the case of gain, the absolute acceleration estimated follows the measured acceleration at above 1 Hz, and in the case of phase at above 10 Hz. The estimated absolute acceleration gain has the form of a high-pass filter as intended. The estimated absolute velocity and position derived by integrating the absolute acceleration have the form of a band filter. The phase differences between them and the estimated absolute acceleration exceed $-90°$ from 10 Hz. The gain of

Table 5.5 Parameters and control gains [2]

Parameter	Value (unit)	Parameter	Value (unit)
μ_0	$4\pi \times 10^{-7}$	R	2.4 (Ω)
N	220 (Turn)	z_0	0.01 (m)
A (magnet)	0.011 (m^2)	i_0	17 (A)
T_1	0.16	k_{zpp}	33
T_2	0.008	k_{zp}	495
T_3	0.25	k_z	0
T_4	0.000176	k_{gp}	26400
T_5	0.016	k_g	0
V_1	1.28	V_2	1.0
V_3	0.2		

Fig. 5.53 Bode diagram for the dynamic filter [2]

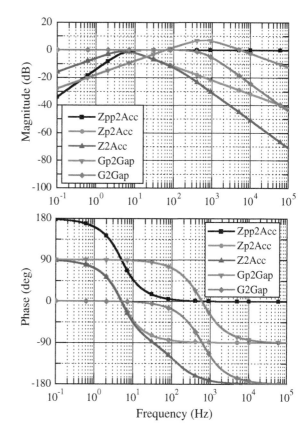

the estimated airgap has the form of a low-pass filter going down from above 40 Hz. These characteristics must be optimized through the choice of gain and parameters in the control loop and the properties of the suspension system and required performance. To verify the design and control of the lift electromagnet, a tester with a

Fig. 5.54 Test rig for
magnetic levitation control

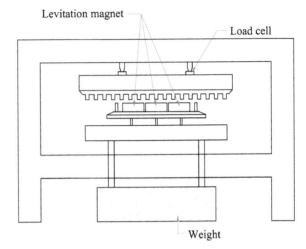

Fig. 5.55 Lift forces
measured and calculated
versus current and airgap [2]

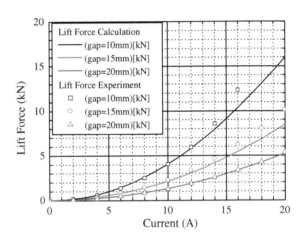

magnet module of 3 magnets was constructed, as shown in Fig. 5.54. The static lift
forces measured are shown in Fig. 5.55 with variances in the current in the magnet
coil and airgap. The lift forces from both tests and simulations are similar, especially at smaller current. It can be confirmed that the required lift force is sufficiently
achieved.

The responses of airgap to ramp reference were measured with 3 different
weights: 860, 1180, and 1500 kg. It can be noted that all three responses well follow
the reference, with a small steady-state error (Fig. 5.56). The error is due to the lack
of integral gain, and thus could be reduced if an integral gain was inserted into the
voltage determination of Eq. (5.56). One of the performance criteria of electromagnetic suspension is the capability of following a deflected guideway. Though
there are several frequencies that must be followed by a lift electromagnet in order

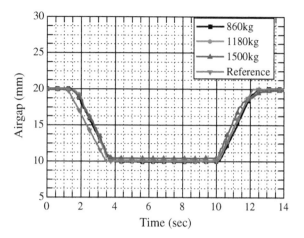

Fig. 5.56 Controlled airgaps following a ramp reference [2]

to avoid physical contact with the guideway, the most important frequency is that which is due to the vehicle passing over spans of the guideway. If the span length is assumed to be 25 m and vehicle speed 550 km/h, the vehicle passing frequency is about 6 Hz. It thus follows that the control bandwidth of the control loop should exceed 6 Hz. The Bode diagram made from these tests is useful for evaluating the control bandwidth. In this test, a sine sweep is used as a reference input to the control loop. Figure 5.57 shows the response of the airgap to a sine sweep input reference. The controlled airgap follows the reference, with little magnitude and phase difference. For the sake of clarity Bode diagrams are drawn from these time responses with simulations. The Bode diagram for 1180 kg presents the control bandwidths of 10 Hz in simulation and 7 Hz in the experiment, as shown in Fig. 5.58. The differences in control bandwidth may be due to practical

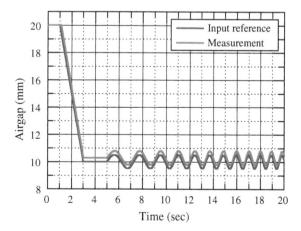

Fig. 5.57 Time response of the magnet for chirp reference [2]

Fig. 5.58 Frequency
response of 1180 kg vertical
load [2]

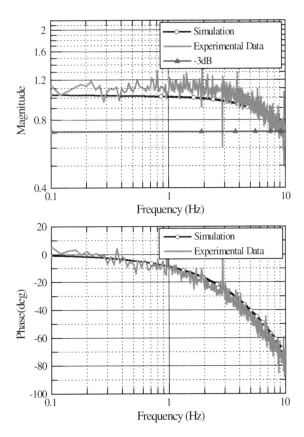

considerations such as mechanical contact between parts. The control gains and parameters determined above could be used as a basis for the experimental vehicle SUMA550.

5.5.7   *Multivariable Control*

Since the maglev vehicle is being supported and guided by electromagnets, the movements of the vehicle can effectively be controlled in three translational and three rotational motions. However, because the movement amplitudes of each direction are absolutely and relatively different, a suitable choice of degrees of freedom in developing the dynamic model is essential for efficient subsequent works. The selection may be dictated by considering the performances to be evaluated and degrees of freedom and external disturbances that effectively influence the performances. In developing the framework for dynamic analysis and controller design of the system under study, the reference frame in Fig. 5.59 may

Fig. 5.59 Vehicle frame of reference

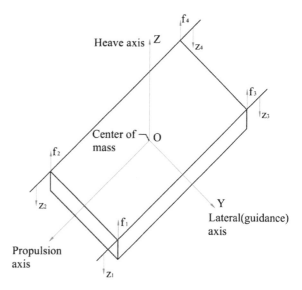

first be introduced. The equations of motion about the center of gravity may be expressed as

$$\left.\begin{array}{l} M\ddot{x} = F_x \\ M\ddot{y} = F_y \\ M\ddot{z} = F_z \\ M_x = I_{xx}\dot{\omega}_x + (I_{zz} - I_{yy})\omega_y\omega_x \\ M_y = I_{yy}\dot{\omega}_y + (I_{xx} - I_{zz})\omega_z\omega_x \\ M_z = I_{zz}\dot{\omega}_z + (I_{yy} - I_{xx})\omega_x\omega_y \end{array}\right\} \quad (5.62)$$

where the 3 angular velocities are expressed in terms of Euler angles.

$$\omega_x = \dot{\phi} + \dot{\psi}\sin\theta$$
$$\omega_y = \dot{\theta}\cos\theta - \dot{\psi}\sin\theta\cos\theta$$
$$\omega_z = \dot{\theta}\sin\theta + \dot{\psi}\cos\phi\cos\theta$$

If the rotations are assumed to be very small, then the angular accelerations may be determined by

$$\left.\begin{array}{l} \dot{\omega}_x = \ddot{\phi} + \left[\ddot{\psi}\theta + \dot{\psi}\dot{\theta}\right] \\ \dot{\omega}_y = \ddot{\theta} + \left[\dot{\psi}\dot{\theta}\phi\theta - \ddot{\psi}\phi - \dot{\psi}\dot{\phi} - \dot{\theta}\dot{\phi}\phi\right] \\ \dot{\omega}_z = \ddot{\psi} + \left[\ddot{\theta}\phi + \dot{\theta}\dot{\phi} - \dot{\psi}\dot{\phi}\phi - \dot{\psi}\dot{\theta}\theta\right] \end{array}\right\} \quad (5.63)$$

Because of the nonlinearity of Eq. (5.63), the numerical integration methods such as
the Runge-Kutta method in Chap. 2 may have to be employed. Although the
levitated vehicle has 6 degrees of freedom, assuming a very small yaw angle and
constant longitudinal velocity, the first order approximations to Eq. (5.62) may be
derived for small perturbations around its nominal operating positions.

$$\left.\begin{aligned} F_z &= M\ddot{z} \\ T_{roll} &= I_{zz}\ddot{\phi} \\ T_{pitch} &= I_{yy}\ddot{\theta} \end{aligned}\right\} \tag{5.64}$$

$$\left.\begin{aligned} F_z &= f_1 + f_2 + f_3 + f_4 \\ T_{roll} &= (-f_1 + f_2 - f_3 + f_4)B \\ T_{pitch} &= (f_1 + f_2 - f_3 - f_4)L \end{aligned}\right\} \tag{5.65}$$

Where for very small perturbations, the heave (z), roll (ϕ) and pitch (θ) can be
determined by the four airgaps, as follows.

$$\left.\begin{aligned} z &= \tfrac{1}{4}(z_1 + z_2 + z_3 + z_4) \\ \phi &= \tfrac{1}{2b}(-z_1 + z_2 - z_3 + z_4) \\ \theta &= \tfrac{1}{2l}(z_1 + z_2 - z_3 - z_4) \end{aligned}\right\} \tag{5.66}$$

L and B (l and b) are the centers of magnets along the length and width of the
vehicle, and M is the total mass of the levitated vehicle. Although the deflections of
the vehicle body are practically present, in the preliminary analysis they may be
neglected. The forces of each magnet are influenced by the currents and airgaps in
other magnets, and as such the cross-coupling effects need to be considered. The
attractive force produced by jth magnet may be expressed by

$$f_j(t) = -k_{ij}i_j(t) + k_{zj}z_j(t) + f_{dj}(t) \tag{5.67}$$

$$\frac{di_j}{dt} = \frac{k_{zj}}{k_{ij}}\frac{dz_j(t)}{dt} - \frac{R_j}{L_j}i_j(t) + \frac{1}{L_j}v_j(t) \tag{5.68}$$

Owing to the cross-coupling effects involved in the above equations, one change in
magnet force influences all other airgaps. Thus, this is to be a four-input four-output
system. If the cross-coupling effects must not be neglected, they need to be considered in the levitation control loop. In the simplest case, where the vehicle's
center of mass coincides with the geometric center, the non-interacting dynamics
represented by Eqs. (5.64) and (5.65) enable each degree of freedom z, ϕ and θ to
be controlled independently by working with the appropriate linear combinations of
f_j. For example, to control the vertical position z, the mean of z_j may be used instead
of each f_j. Using these relations, the control loop can be made ideally
non-interacting and it is possible to design for each degree of freedom without the
mutual cross-coupling. The control systems using this scheme may be based on

Eqs. (5.64) and (5.66). This control method has been given the term "integrated control." An alternative design method is to consider the cross-coupling involved in Eqs. (5.65) and (5.66) and to exert control based on a tight association between a magnet and its nearest airgap sensor, i.e. the force f_j is controlled by the airgap z_j. This control scheme has been given the term "local control," and it is possible to design each magnet-airgap sensor loop separately. Because of its ability to control each degree of freedom independently, the main advantage of local control is the applicability of well-developed control theories. One of the ways of improving the usability of local control is to weaken the cross-coupling by introducing appropriate geometrical relations and suspension elements, the effect of each levitation force on other degrees of freedom being minimized. Despite such minimized cross-coupling, multi-variable control may be needed. A multivariable control scheme to reduce the variations on roll, pitch, and yaw motion is introduced through case studies.

- **Roll control**: Though the experimental vehicle SUMA550 in Sect. 5.3 is controlled by using local control method, a roll control of local control was implemented here to demonstate the effectivess of the multivariable control concept [3]. Roll motions of the vehicle may occur due to the uneven height of the test track. To reduce the amplitude of roll angle, a scheme to modify the control voltages in the magnet coils on both sides, i.e. left and right side, is demonstrated. The roll angle is first calculated by sharing the airgap signals from the other side. The roll and its time derivative are derived as

$$\widehat{g_r} = \Delta c_{right}(t) - \Delta c_{left}(t)$$
$$\dot{\widehat{g_r}} = \Delta \dot{c}_{right}(t) - \Delta \dot{c}_{left}(t)$$

The control voltage related to roll angle is determined by multiplication of the roll control gains.

$$v_{roll}(t) = k_{rgv}\widehat{g_r} + k_{rg}\dot{\widehat{g_r}} \tag{5.69}$$

Then, this control voltage is added to that from the state feedback law in Eq. (5.56).

$$\Delta v(t) = k_{zpp}\Delta\ddot{z}(t) + k_{zp}\Delta\dot{z}(t) + k_z\Delta z(t) + k_{gp}\Delta\dot{c}(t)$$
$$+ k_g\Delta c(t) + v_{roll}(t) \tag{5.70}$$

The configuration to implement the roll control described above is shown in Fig. 5.60. The roll control gains are chosen to be $k_{rgv} = 80$, $k_{rg} = 15$ through an iterative design procedure. To evaluate the effectiveness of the roll control, some controlled experiments were conducted with the SUMA550. To intentionally induce the roll motion of the vehicle being levitated, the airgap references for the magnets on the right side are excited from the nominal value of 10 mm with a sine wave. The airgaps, representing roll angles, from with and without the roll control are compared in Fig. 5.61, where the relative airgaps between both sides

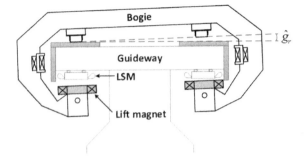

Fig. 5.60 Concept of roll control scheme

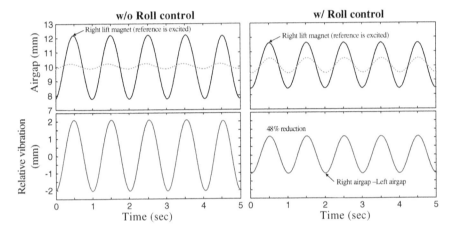

Fig. 5.61 Effects of roll control on relative airgaps between both side magnets

are largely reduced through roll control. It has been observed that a reduction in the structural vibration of bogie frame has also been exhibited. Even though the roll control appeared very much to have positive effects, it needs to be carefully incorporated into the maglev system due to the airgap deviation restrictions, which may be a more fundamental requirement than the roll motion (Fig. 5.62).

- **Pitch control**: The pitch control for the bogie in the SUMA550 may be almost similar to that of the roll control [3]. To reduce the amplitude of pitch angle, a scheme to modify the control voltages in the magnet coils on both sides, i.e. front and rear, is demonstrated. First, by sharing the states in the front and rear magnets, the five internal states are calculated and then the control voltage to reduce the pitch is derived as

$$v_{pitch}(t) = k_{pza}\dot{\widehat{g}}_p + k_{pg}\widehat{g}_p$$

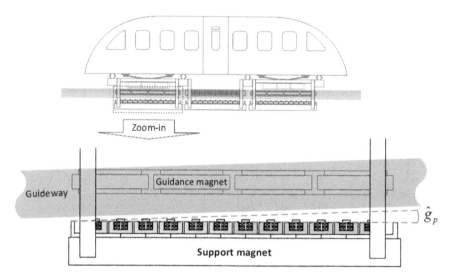

Fig. 5.62 Concept of pitch control [3]

where

$\widehat{\dot{g}}_p$ relative airgap velocity

$\widehat{g}_p$ relative airgap

And $k_{pgv} = 40$ and $k_{pg} = 6$. Consequently, the control voltage for each magnet may be modified as

$$\Delta v(t) = k_{zpp}\Delta\ddot{z}(t) + k_{zp}\Delta\dot{z}(t) + k_z\Delta z(t) + k_{gp}\Delta\dot{c}(t)$$
$$+ k_g\Delta c(t) + v_{pitch}(t)$$

To assess the effectiveness of the pitch control, some controlled experiments have been conducted with the SUMA550. Similarly to the roll control, the frontal magnets of the levitated vehicle at a nominal position are excited by a square wave of 1 Hz assuming steps at rail joints, intentionally reducing the pitch motion. The influences of the pitch control on reducing pitch can be well observed in Fig. 5.63, which indicates that the pitch was dramatically reduced thanks to the pitch control. The comments on roll control above may be equally applied to this pitch control.

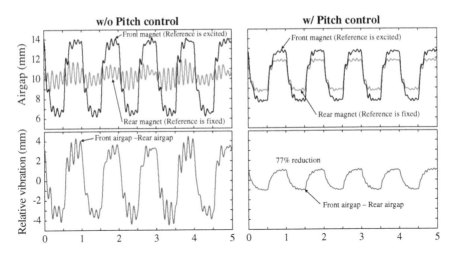

Fig. 5.63 Effect of the pitch control on relative airgaps between front and rear magnets [3]

5.5.8 Ride Control

As indicated in the previous sections, a feedback control law for an electromagnetic attraction system must usually be designed to meet the suspension stability and ride comfort requirements. Since the suspension stability is closely related to the physical dimensions of a magnet, a suspension system design derived only from ride comfort considerations may not provide a sufficient margin of stability. From a ride quality viewpoint, the dynamic responses of the vehicle suspensions at frequencies in the range of 1–2 Hz are the important design parameters. This aspect suggests that the natural physical frequency of the passenger vehicle suspension needs to be in this range of about 1 Hz. In high-speed vehicles, a low natural frequency implies a relatively large magnet-guideway clearance. For superconducting systems, such large airgap variations may be accommodated. Thus the primary suspension of superconducting systems may be designed to have a natural frequency of 1–2 Hz. On the other hand, the electromagnetic attraction system may have difficulty accommodating such a large clearance, because a larger clearance requires a larger magnet and lower lift force/input power. Since stability is the most important design consideration in electromagnetic attraction systems, the design of a stabilizing controller is related to allowable steady-state error. This in turn is related to the suspension stiffness, i.e. the closed-loop natural frequency of the system. Consequently, the frequency of the electromagnetic systems is restricted. For full-scale systems using typical electromagnetic suspension, their natural frequencies are known to be 4–10 Hz with an airgap of 10–15 mm. Thus, any secondary suspension between the passenger cabin and the levitation magnet may be introduced to the system to achieve the required primary suspension stiffness while allowing an acceptable ride quality. Typical secondary suspension systems consist of a spring and a damper. The configuration is conceptually illustrated in Fig. 5.64.

Fig. 5.64 Configuration of the suspension system with a secondary suspension

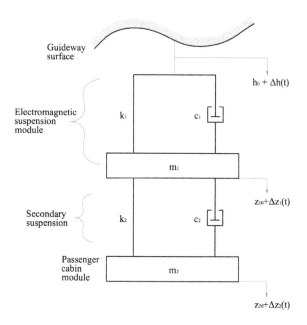

The suspension design with acceptable ride comfort may be generated through a two-stage procedure. In the first stage, the suspension is stabilized to derive the adequate frequency response and guideway following properties. That is, the stiffness, natural frequency and damping ratio of the primary suspension means, i.e. magnet, are chosen. Using these chosen values, the variations in airgap and acceleration levels are predicted with a given guideway surface roughness represented by PSD. Through this stage, the levitation magnet and its associated driver are selected. The next stage is concerned with the design of secondary suspension to improve the overall system performance. The stiffness and damping of the secondary suspension system are adjusted to restrict the acceleration levels and airgap deviations through an interactive design procedure. In practice, the control gains may be fine-tuned during test runs over the track. As well, since the physical secondary suspension may be difficult to modify, the fine-tuning is mainly related to the control gains as numerical values, based on the goal of achieving both ride comfort and suspension stability.

5.5.9 Measurement

The contactless nature of the operation of maglev systems demands non-contact gap sensors and accelerometers. As in most control systems, hardware and software filters are also needed to remove the noise involved in measured signals. This section provides some criterion for the selection of adequate sensors. The criteria for judging the sensors include the bandwidth, linearity over operating range, robustness and stability under all operating conditions, and immunity from noise, radiation and stray magnetic fields. The meanings of these terms and the related considerations are briefly summarized as follows.

- **Bandwidth**: Bandwidth is a measurable frequency range of a sensor, and it is preferred to be 100–500 times the control bandwidth of the system under control.
- **Operating range**: As the range to be measured with the sensor, it is chosen considering the range of the change in the physical quantity. If the operating range is small, the resolution becomes better, but easily approaches the saturation region. On the other hand, if the range is wider, saturation may be difficult to reach and the resolution becomes poorer.
- **Linearity**: The linearity assures identical characteristics over the operating range. If necessary, an additional compensatory circuit may be incorporated.
- **Noise**: The noise included in a measured signal may be caused by thermal effects and high-frequency switching of electronics components and signal processing.
- **Magnetic field**: Some sensors are influenced by external magnetic fields. Particularly in maglev systems, the magnetic field effects must be considered in order to choose appropriate sensors.
- **Gap sensor**: Of the various types of contactless gap sensors, inductive sensors are the type most frequently used in maglev systems. The reason for this is that the steel reaction surface can be used as a target for the gap sensors. The principle is based on eddy current induced in the target. The sensor consists of two coils, in which one coil is excited by AC current and the other is used to measure the induced eddy currents. The commercially available gap sensors have a bandwidth of 500 Hz–20 kHz; in a maglev system, sensors with a bandwidth of over 1 kHz may be chosen considering the control bandwidth of around 100 Hz in most cases. For operating range, 0–25 mm may be an adequate choice because most operate within 20 mm airgaps. Some inductive type sensors essentially have a non-linear characteristic. This may be overcome through an embedded compensator circuit.
- **Accelerometer**: The sensing element of an accelerometer consists of a spring-mass system that deflects when subjected to acceleration. The deflection of the seismic mass is a linear function of acceleration over the operating range, which is constrained by the natural frequency and damping ration of the seismic system. Because of the bandwidth, accuracy and linearity requirements, not all accelerometers can be applied to feedback control systems. Of the various accelerometer types, piezoelectric and MEMS (Micro Electric Mechanical System) type are used most frequently. For ride comfort, accelerations at 0–30 Hz are of importance, and thus a bandwidth of 150 Hz may be adequate allowing for the phase lag. In practice, though the steady-state acceleration level is not expected to be higher than 0.1 g, the transient acceleration can be as high as 3 g. Consequently, the operating range of ±3 g may be selected.

Fig. 5.65 Circuit diagram of a two-quadrant chopper

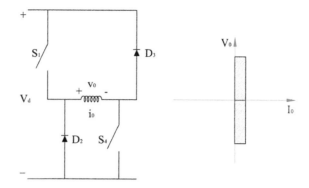

5.5.10 Electronics

One of the factors in the viability of the electromagnetic type systems relies upon the current carrying and switching capacity to produce controlled magnetic levitation systems capable of supporting load. The levitation force is only derived by the magnitude of current in the magnet coil, regardless of the direction of electric current flow. Thus, one direction of current flow is used in normal electromagnets. For this reason, a two-quadrant (two phase) chopper configuration, i.e. a driver, with reduced electronic components may be used to excite the electromagnets. The driver circuit used for the SUMA550 is schematically shown in Fig. 5.65. If two choppers S_1, S_4 are switched on, the voltage at $v_0 = V_0$ flows to the magnet in the positive direction; in contrast, if they are switched off and there are residual currents in the positive direction, the voltage becomes $v_0 = -V_0$ because of current flowing across diodes D_2 and D_3. At this time, since I_0 flows in the positive direction and V_0 in the negative direction, the power flows inversely. The direction of V_0 depends on the duration for which power is switched on or off. However, the inverse flow of current is not allowed for because the current flowing across D_2 and D_3 cannot be controlled. One of the most important aspects in the power amplifier for the electromagnets is to maximize the rise and fall rates of the currents in the magnet coil. One of the factors to determine the control bandwidth of the magnet is these two rates. Of course, rapid and identical rates are naturally recommendable. One of the ways to increase the two rates is to use high source voltage.

5.6 Guidance Control

In addition to levitation force, guidance forces are needed in all the maglev vehicles running over a track. For U-shaped magnets, some guidance forces are provided by the fringe effect of the lift magnet. That is, one U-shaped magnet offers both levitation and guidance forces. However, if the running speed increases or lateral clearance needs to be strictly limited, the guidance force requirement becomes more important. Typically, one of two methods could be used to obtain the required guidance force. The first scheme is to achieve the two forces with the same magnet, the guidance force being provided passively. The improved configuration from the scheme to increase guidance force is a staggered configuration, in which two magnets are oppositely offset from the common axis by an equal distance. Though this staggered configuration is also a simple method, due to its passive nature, a feedback control may be incorporated when the guidance force requirements are severe. Even though this method is rather complex in terms of its configuration and implementation, the added active nature has advantages in high-speed or high precision systems. This section presents three methods for achieving guidance force.

5.6.1 Combined Magnets

As indicated above, a U-core magnet can provide levitation and some guidance forces. Figure 5.66 shows the general configuration of a U-core magnet facing a U-shaped reaction rail with a relative displacement d. For this configuration, the levitation and guidance forces may be derived. The lateral guidance forces $\overline{F}_{1y}, \overline{F}_{2y}$ are generated by the "fringing effect" described in Chap. 2. If it is assumed that the airgap along the whole magnet's length is identical and the fringe flux is the same as the airgap flux, then the resulting levitation and guidance for one magnet having two poles are expressed as [1]

$$
\begin{aligned}
F_z &= \overline{F}_{1z} + \overline{F}_{2z} \\
&= F_0 \left[1 + \frac{z}{\pi w_m} + \frac{\beta - d}{\pi w_m} \tan^{-1}\left(\frac{z}{\beta - d} \right) + \frac{\beta + d}{\pi w_m} \tan^{-1}\left(\frac{z}{\beta + d} \right) \right]
\end{aligned}
\tag{5.71}
$$

$$
\begin{aligned}
F_y &= \overline{F}_{1y} - \overline{F}_{2y} \\
&= F_0 \left(\frac{z}{\pi w_m} \right) \left[\tan^{-1}\left(\frac{\beta - d}{z} \right) - \tan^{-1}\left(\frac{\beta + d}{z} \right) \right]
\end{aligned}
\tag{5.72}
$$

where

$$
F_0 = 2F = \frac{B^2}{2\mu_0} (2 w_m l_m) \rightarrow \text{ideal lift force per magnet}
$$

It is relevant here to derive the ratio of lift to guidance forces. For this ratio, the figures of the magnet for HSST are quoted here [4]. The main specifications are listed in Table 5.6. In addition, the measured levitation force and guidance/levitation forces are given in Figs. 5.67 and 5.68, respectively. For HSST, the required guidance force is 20 % of the vehicle weight. From Fig. 5.68, it is

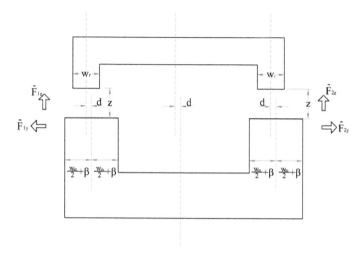

Fig. 5.66 Displaced magnet geometry

Table 5.6 Specifications of lift magnet [4]

Parameters	Values
Suspended mass: m	2600 kg
Pole area: A	0.0345 m^2
Coil turn: N	304
Coil current: I	29 A
Airgap: z	0.008 m
Vacuum permeability: μ_0	$4\pi \times 10^{-7}$
Core pole width	0.028 m

Fig. 5.67 Levitation force-airgap characteristics of the HSST levitation module [4]

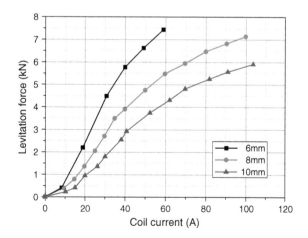

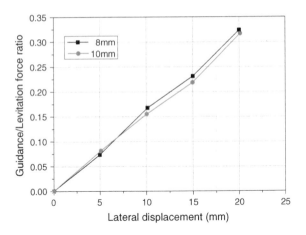

Fig. 5.68 Guidance/Levitation force ratio of the HSST magnet [4]

shown that lateral displacement of the magnet within 15 mm satisfies the guidance force requirement. Based on these results, an empirical relationship of the guidance/levitation forces was derived as

$$\frac{f_l}{f_v} = 0.43 \frac{d}{wp}$$

where f_l is the guidance force, f_v the guidance force, d the lateral displacement, and wp the magnet core pole width. The coefficient of 0.43 was chosen based on the measurements. Here, the factor to note is the saturation effect. The saturation effect is demonstrated in Fig. 5.67, with the forces approaching maximum values with increasing currents. This saturation effect implies that the guidance force also has a saturation effect, and thus the magnet must be operated within the unsaturated region.

The analytical force expressions above suggest that if two limbs of a U-core magnet can be controlled separately, then guidance control may be possible, even through the same magnet. The lateral forces in a single magnet can be produced according to the difference in two airgap flux densities. That is, adjusting the flux density difference in both poles allows the guidance force to be controlled. If B_l and B_r are varied such that $B_l^2 + B_r^2$ remains unchanged, then the guidance force can be controlled while the total lift force is constant. Though this guidance control scheme is conceptually simple to implement, its usability is limited by the constraint on the constant lift force.

5.6.2 Staggered U-core Magnets

To partially overcome the limitations of using a single magnet for guidance force, the staggered configuration, known as Krauss-Maffei configuration, may be used without a feedback control [1]. This configuration is schematically illustrated in Fig. 5.69. Because of the symmetrical arrangement of each staggered pair of magnets, the position of the suspension system in steady-state will coincide with the axis of the U-shaped rail. The main advantage is the continuous control of both levitation and guidance forces. In steady-state, the nominal lift force can be controlled by adjusting the sum of currents in two magnet coils with the measured airgap. In contrast, the nominal lateral force is derived by the difference between the two magnet currents. To quantify the staggering effect, the relationship between the main dimensions and the lift and lateral forces needs to be established. For analytical convenience, the pole widths of the magnet and rail are assumed to be identical. Then, for lateral displacement d, from Fig. 5.70,

$$\text{Magnet A: } \beta_l = \Delta - d \quad \text{and} \quad \beta_r = -(\Delta - d) \tag{5.73}$$

$$\text{Magnet B: } \beta_l = \Delta + d \quad \text{and} \quad \beta_r = -(\Delta + d) \tag{5.74}$$

Combining Eqs. (5.71)–(5.74)

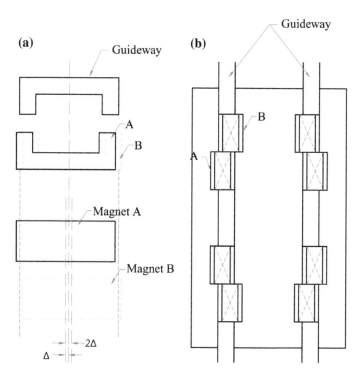

Fig. 5.69 Lift/Guidance through staggered U-core magnets

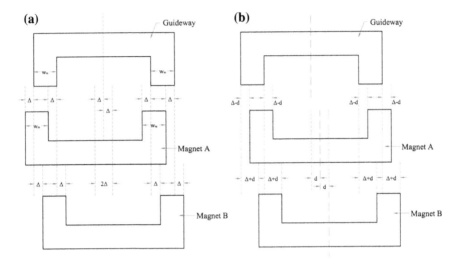

Fig. 5.70 Configuration of the magnet guideway **a** without any displacement and **b** with a displacement d to the right

$$F_{za} = \overline{F}_{1za} + \overline{F}_{2za} = F_0 \left[1 + \frac{z}{\pi w_m} + \frac{\Delta - d}{\pi w_m} \tan^{-1} \left(\frac{z}{\Delta - d} \right) \right. $$
$$\left. - \frac{\Delta - d}{\pi w_m} \tan^{-1} \left(-\frac{z}{\Delta - d} \right) \right] \qquad (5.75)$$

$$= F_0 \left[1 + \frac{z}{\pi w_m} + 2 \frac{\Delta - d}{\pi w_m} \tan^{-1} \left(\frac{z}{\Delta - d} \right) \right] \qquad (5.76)$$

$$F_{zb} = \overline{F}_{1zb} + \overline{F}_{2zb}$$
$$= F_0 \left[1 + \frac{z}{\pi w_m} + 2 \frac{\Delta + d}{\pi w_m} \tan^{-1} \left(\frac{z}{\Delta + d} \right) \right] \qquad (5.77)$$

$$F_{yb} = \overline{F}_{1yb} - \overline{F}_{2yb}$$
$$= F_0 \frac{z}{\pi w_m} \left[\tan^{-1} \left(\frac{\Delta - d}{z} \right) - \tan^{-1} \left(-\frac{\Delta - d}{z} \right) \right] \qquad (5.78)$$
$$= 2F_0 \left(\frac{z}{\pi w_m} \right) \tan^{-1} \left(\frac{\Delta - d}{z} \right)$$

$$F_{yb} = \overline{F}_{1yb} - \overline{F}_{1yb} \qquad (5.79)$$

Consequently, the total lift force and net lateral force are derived as

$$F_z = F_{za} + F_{zb}$$
$$= 2F_0 \left[1 + \frac{z}{\pi w_m} + \frac{\Delta - d}{\pi w_m} \tan^{-1} \left(\frac{z}{\Delta - d} \right) + \frac{\Delta + d}{\pi w_m} \tan^{-1} \left(\frac{z}{\Delta + d} \right) \right] \quad (5.80)$$

$$F_y = F_{ya} - F_{yb} = 2F_0 \left(\frac{z}{\pi w_m} \right) \left[\tan^{-1} \left(\frac{\Delta - d}{z} \right) - \tan^{-1} \left(\frac{\Delta + d}{z} \right) \right] \quad (5.81)$$

The above equations indicate that the staggered separation could have a strong effect on the slope of lift force. Since this levitation and guidance force control concept is based on the two limbs control of one U-core magnet, the levitation and guidance force could be changed by controlling the sum and difference of the current in a pair of magnets.

5.6.3 Separate Magnets

Though the two methods of obtaining guidance force that were introduced in the previous sections have a simple configuration, their ability to generate guidance forces is limited due to there being no magnet dedicated to providing guidance force only. Thus, the introduction of a dedicated magnet for guidance may be needed, particularly in high-speed or high precision operations. For guidance control, one DOF model is introduced, as shown in Fig. 5.71. The equations of the guidance force and current are identical, as in the levitation control. That is,

$$F(i, y) = \frac{\mu_0 N^2 A}{4} \left(\frac{i}{y} \right)^2 = K \left(\frac{i}{y} \right)^2 \quad (5.82)$$

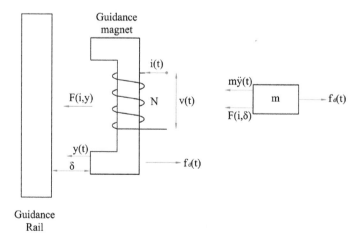

Fig. 5.71 1 DOF dynamic guidance model

The equation of motion for the mass is expressed as

$$m\ddot{y} + F(i, y) - f_d = 0 \tag{5.83}$$

The state equations can be derived by modifying the equations for levitation.

$$\begin{bmatrix} \Delta\dot{\delta} \\ \Delta\ddot{y} \\ \Delta\dot{i} \end{bmatrix} = \begin{bmatrix} 0 & 1 & 0 \\ \frac{k_y}{m} & 0 & -\frac{k_i}{m} \\ 0 & \frac{i_0}{y_0} & -\frac{R}{L_0} \end{bmatrix} \begin{bmatrix} \Delta\delta \\ \Delta\dot{y} \\ \Delta i \end{bmatrix} + \begin{bmatrix} 0 \\ 0 \\ \frac{1}{L_0} \end{bmatrix} [\Delta v] + \begin{bmatrix} 0 \\ \frac{1}{m} \\ 0 \end{bmatrix} [f_d] \tag{5.84}$$

The constants are represented identically to those of the levitation control. With these equations, the transfer function between the current and lateral displacement is derived as

$$H(s) = \frac{\Delta Y(s)}{\Delta I(s)} = \frac{k_i}{ms^2 - k_y}$$

where

$$m\Delta\ddot{y} = k_y \Delta\delta + k_i \Delta i$$

The transfer function indicates that this system is unstable due the one pole on the right half plane, and thus a feedback loop is needed to stabilize the system. The state feedback scheme in Sect. 5.6 could be applied. The configuration of the guidance control is given in Fig. 5.72. The same control law in Sect. 5.6 can be used, after modifying the control gains for guidance control.

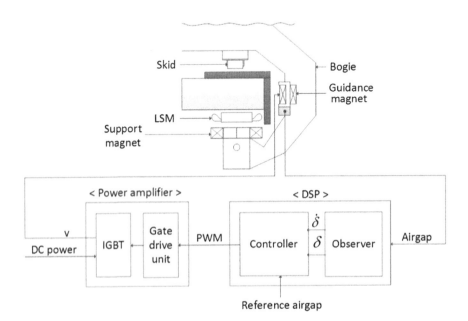

Fig. 5.72 Configuration of the guidance control with the states feedback

Fig. 5.73 Step responses of the guidance control system

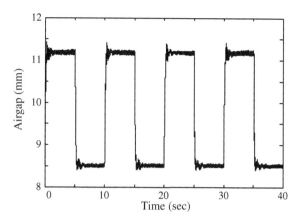

To demonstrate the guidance control effect, the scheme described above is applied to the experimental vehicle SUMA550. The control gains are selected as $k_{zpp} = 100, k_{zp} = 2000, k_z = 0, k_{gp} = 2000$, and $k_g = 3$. The step responses in one corner are measured as shown in Fig. 5.73, and show the ability to follow the reference with small errors. The frequency responses obtained from the step response experiments are analyzed in Fig. 5.74, with the control bandwidth being 2.5 Hz.

To assess the guidance control effect while running over the 150 m long test track, the running tests with and without full load were undertaken. The specified lateral clearance deviation is 3 mm. Looking at Fig. 5.75, it can be noted that the

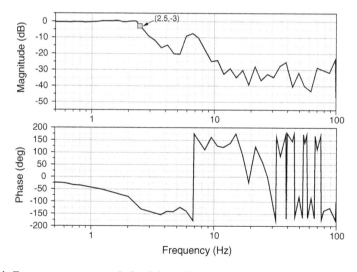

Fig. 5.74 Frequency response analysis of the guidance system

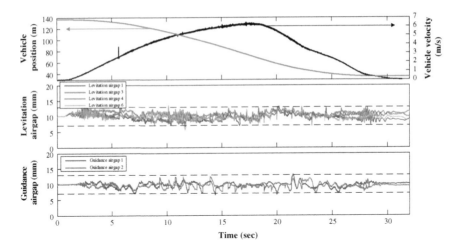

Fig. 5.75 Effects of guidance control on clearance between guideway and guidance magnet

guidance clearances were within 3 mm, satisfying the clearance deviation requirement.

The guidance control based on a multivariable control scheme could be applied considering the yaw motion of the bogie in the vehicle. The concept is basically the same as those of the pitch and roll control (Fig. 5.76). The main idea is to add the additional voltage derived from the yaw motion to each control voltage determined by the state feedback control law for each. The yaw motion can be calculated by sharing the states of each corner (Fig. 5.77). Then, the control voltage to reduce the yaw motion can be derived by

$$v_{yaw}(t) = k_{pya}\widehat{\ddot{z}}_r + k_{pyv}\widehat{\dot{z}}_r + k_{py}\widehat{z}_r + k_{pgv}\widehat{\dot{g}}_r + k_{pg}\widehat{g}_r \qquad (5.85)$$

where
$\widehat{g}_r$ relative lateral position (airgap)
$\widehat{\dot{g}}_r$ relative lateral velocity
$\widehat{z}_r$ difference in absolute lateral position
$\widehat{\dot{z}}_r$ difference in absolute lateral velocity
$\widehat{\ddot{z}}_r$ difference in absolute lateral acceleration

The resulting guidance control voltage for each corner is derived as

$$\begin{aligned} \Delta v(t) = k_{ypp}\Delta\ddot{y}(t) + k_{yp}\Delta\dot{y}(t) + k_y\Delta y(t) + k_{gp}\Delta\dot{c}(t) \\ + k_g\Delta c(t) + v_{yaw}(t) \end{aligned} \qquad (5.86)$$

Where the control gains are chosen as $k_{pya} = 40$, $k_{pyv} = 400$, $k_{py} = 0$, $k_{pgv} = 400$, and $k_{pg} = 7.5$. To evaluate the guidance control considering the bogie yaw

Fig. 5.76 Yaw angle of
bogie in SUMA

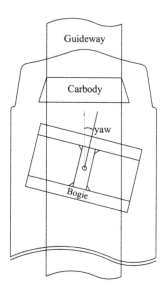

Fig. 5.77 Concept of yaw
control

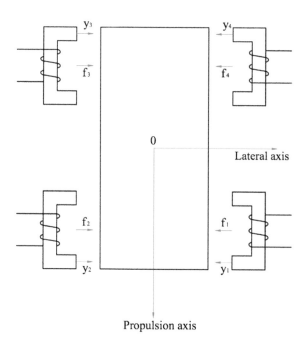

motion, some controlled experiments are undertaken. To intentionally induce the
yaw motion, one corner magnet in front was excited based on the reference with
sine and square waveforms. The airgap responses are given in Figs. 5.78 and 5.79,
which show the reduction in the yaw (relative position) with the yaw control.

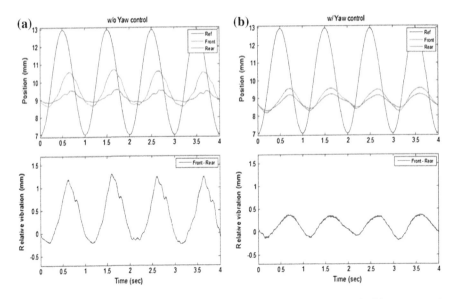

Fig. 5.78 Sinusoidal responses of the guidance magnets at 1 Hz without and with yaw control: **a** without yaw control, **b** with yaw control

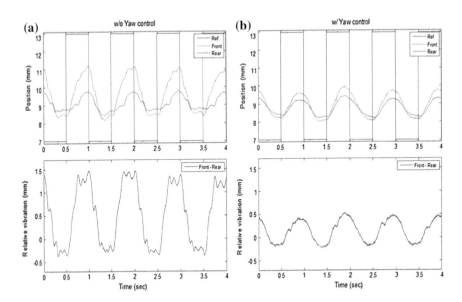

Fig. 5.79 Square responses of the guidance magnet at 1 Hz without and with yaw control: **a** without control, **b** with control

5.7 Vehicle/Guideway Interaction

5.7.1 Issues

The construction cost of the guideway has been known to exceed 60–70 % of the total initial investment in a maglev train. Due to this large proportion of the cost, a more slender guideway and looser constructional tolerances are desperately demanded in order to reduce the overall construction costs. A lighter guideway has increased flexibility, resulting in a tight coupling needed between the control loops in the vehicle and the flexible guideway. Due to this strong coupling, some vibration problems may appear. If these vibrations are not sufficiently controlled, then a divergence may occur. Loosening the guideway tolerances causes an increase in airgap variation, which may lead to mechanical contact between the vehicle and guideway, which in turn brings impulsive high accelerations, and may destabilize the levitation control system. For cost effectiveness, the goal is to loosen the guideway tolerance while maintaining stability. A compromise between stability and loosening tolerances can be achieved through an analysis of vehicle/guideway dynamic interaction. Here, it is relevant to summarize the issues related to dynamic interaction [5–8].

- **Guideway deflection**: In elevated guideway designs, the deflection due to a moving vehicle must be limited by allowable vertical acceleration levels and airgap variations. For a high speed vehicle, the deflection on the mid span may be chosen to be $\delta = l/4000$ mm (l: span lenth(m)). For a low-speed vehicle, $\delta = l/1500 \sim 2000$ mm may be selected. The final deflection limit depends on the particular system.
- **Span length** (l): The span length may be derived by considering the structural parameters. One of them is the vehicle passing frequency over spans ($f = (vehicle\ speed/l)$Hz). This frequency may bring about resonance with the primary or secondary suspensions. The vehicle speed at $f = 1$ Hz is not undesirable for ride comfort. Sufficient separation between the vehicle passing frequency and the natural frequency of the primary suspension is more important from the perspective of stability. For 500 km/h over a span length of 30 m giving a 5 Hz of vehicle passing frequency, a natural frequency above 10 Hz may be recommended.
- **Mass of guideway**: To improve the stability, the guideway having a larger weight compared to vehicle mass has a positive effect. For this reason, concrete beams are preferred.
- **Natural frequency of main beam**: The natural frequency is directly determined by the guideway mass and stiffness. Considering the natural frequency only, a sufficient separation between the fundamental natural frequency and the primary natural frequency is desirable to avoid resonance. A steel beam with a shorter length may have a higher natural frequency that is closer to the primary suspension frequency, which may induce a vibration problem. For example, this

type of problem could be caused by route switches made of steel. This vibration may be more severe at very low running speeds than at high-speeds.

- **Natural frequency of track**: The track usually consists of sleepers and rails. Its natural frequencies may be higher at 40–100 Hz, which is above the control bandwidth of the primary suspension. However, sometimes these high frequency vibration modes may induce high frequency vibrations and acoustic noises. A control scheme with a phase advancer with 2nd or 3rd time derivative of airgap may be incorporated into the control loop.
- **Rail joint**: Rail joints with steps or gaps may bring impulsive airgap signals. For this reason, the rail joints need to be carefully controlled.
- **Surface roughness**: Since the guideway surface roughness affects ride comfort, fine surface roughness will give a better ride quality, but it may be costly. One method for achieving ride comfort is to ignore high-frequency surface irregularity.
- **Long-stator profile**: In LSM propulsion methods, the tooth-shaped stator pole face may induce a variation in lift forces, which may be a vibration exciter.

These issues must be investigated in the design stage and by running tests with the aid of the vehicle/guideway dynamic interaction analysis experimentally or numerically. Consequently, from an economic and running performance view point, a dynamic interaction study may have an important role in the success of maglev vehicles. As a special case, Fig. 5.80 shows a longer span guideway, 50 m long. In practice, the vehicle runs over the suspended longer span guideway because of its lower natural frequency (3 Hz) and heaver weight than the vehicle [6].

Fig. 5.80 EMS type vehicle running over a long span elevated guideway

5.7.2 Analysis Models

The scale and accuracy of the vehicle/guideway dynamic interaction models are chosen considering the objectives and needed efforts. Simplified models may be efficiently used in early design stages, providing parametric studies with less cost. The introduction of control loops into the dynamic models needs to be carefully considered because the control loop could be changed. The most rigorous model, a highly enhanced virtual prototyping model, can provide various dynamic characteristics with fewer simplifications. On the other hand, this model requires much more effort and input data. With the development of the virtual prototyping model, the vehicle/guideway dynamic interaction problem can be more and more effectively investigated.

- **Simplified model**: 1–3 DOF models have been used in the early design stages and in establishing the baseline specifications of vehicle and guideway. The most simplified model represents a control loop as an equivalent spring-damper. The control loop was incorporated into the 1–3 DOF models [6]. Using this model, the issues summarized in Fig. 5.81 can be efficiently evaluated, giving estimates of the vehicle and guideway design parameters.
- **Virtual prototyping model**: This model is the most developed one aiming at replacing the physical prototyping, providing reduced development time and costs [9–11]. The features are as follows:

  - All bodies, joint and force elements are included in the model.
  - Flexibility of bodies can be considered using FE model.
  - Feedback control loops for the electromagnets are included.
  - Positions, velocities, accelerations, forces, deformations as well as control performances are obtainable with a high level of accuracy
  - On the other hand, much more effort in modeling is required.

Since each modeling technique has advantages and disadvantages, it is recommended to use them in a complementary manner depending on the purposes. In the

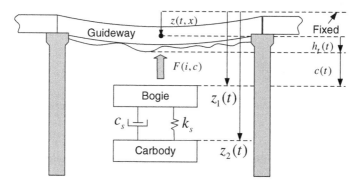

Fig. 5.81 DOF electromagnetic suspension model with a control loop [6]

Fig. 5.82 Dynamic simulation with vehicle/3D FE guideway model: **a** flexible guidewway and **b** flexible bogie models

initial stages, simplified models may be preferred, while in detailed design stages, the virtual prototyping model may be more useful, especially in the analysis of vehicle/guideway dynamic interactions. Figure 5.82 demonstrates the vehicle/

Fig. 5.83 Curving simulation
[10]

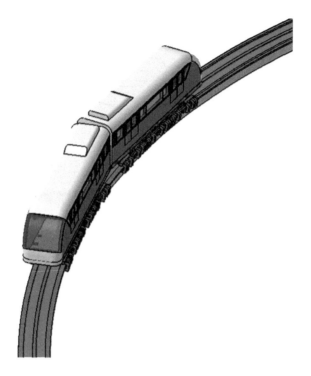

guideway interaction simulation through the virtual prototyping model that includes all bodies and their flexibilities, control loops, and suspension elements. It can be seen that the guideway became deflected due to the moving vehicles. Moreover, a curving simulation in space is possible considering the curved and banked track. With these simulations, the design parameters for vehicle and guideway could be optimized to achieve cost reduction and the performance requirements (Fig. 5.83).

5.8   Emergency Braking

5.8.1   Mechanical Brake

In maglev vehicles, the service brake is provided by the linear motors, i.e. regenerative brake. In case of emergency, very high deceleration must be provided with a safety landing. To fulfill these two requirements, a skid made of sintered alloy is employed. The skid may be placed underneath the bogie frame, as shown in Fig. 5.11a. Brake forces are produced by the friction between the skid and reaction rail. To demonstrate the emergency braking with the skid, the SUMA550 over a

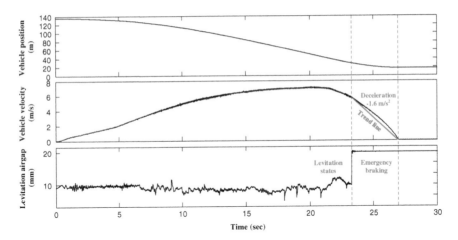

Fig. 5.84 Deceleration during emergency braking

150 m test track was stopped by the skid. As shown by the results in Fig. 5.84, a deceleration of about 1.6 m/s^2 was attained.

5.8.2   Electrical Brake

For high-speed vehicles, the auxiliary electrical brake may be used over a speed range of 10–200 km/h in the event of an emergency. The brake operates based on the eddy current effects induced in a conducting sheet. As described in Chaps. 2 and 3, if an electromagnet excited by a D.C. current moves above a conducting sheet, in practice a steel rail, eddy-currents are induced in it. The induced currents in turn produce magnetic fields. The interaction between the magnetic fields from the magnet and induced eddy-currents generates a drag force and an attraction force. The former is used for stopping. Owing to the operation on eddy current effects, the braking force is dependent on the vehicle speed, number of poles and current of the electromagnet. For demonstration, an electromagnet for braking was designed and implemented in the experimental vehicle SUMA550. The geometry and main dimensions are shown in Fig. 5.85. A aluminum sheet coil with a section area of 0. 3 × 69 mm^2 and 143 turns is used for the magnet. Nominal airgap is set as 10 mm. To evaluate its performance, a 2D model is employed and the results are given in Fig. 5.86. The peak force is achieved at 100 km/h and the force decreases with increasing and decreasing speed.

Fig. 5.85 Analysis model of an eddy-current type brake

Fig. 5.86 Braking
force-speed characteristics

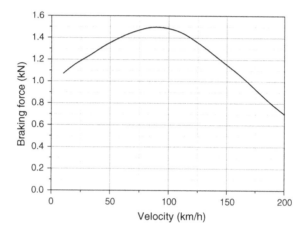

5.9   Route Switching

Magnetically levitated vehicles require their own unique route switching mechanism due to their contactless operation. There is a need for a special lateral guidance
mechanism at the switching point. In particular, in the electromagnetic vehicles
wrapping the guideway, the mechanism may be fairly involved. The desirable
mechanism must minimize the discontinuity of the guiderail at the switching point.
A guideway discontinuity could cause physical contact between the vehicle and
guideway or impulsive airgap signals which profoundly influence the suspension
stabilization control loop. Two types of switching mechanisms are currently in
service. The well-known Transrapid is using a bending type switch, which is shown
in Fig. 5.87, in which the bending is achieved by a hydraulic actuator. It has the
ability to choose one of three possible directions at a junction. The main advantage
is the continuous profile by bending the guideway beam made of steel structure.
During transition through the switch linking with left or right directions, the vehicle
can run at a speed of 100 km, and 500 km/h during operation over a straight, i.e.
normal operation. However, as it uses steel beams, some vibration problems may
occur because of its light weight and high natural frequency.

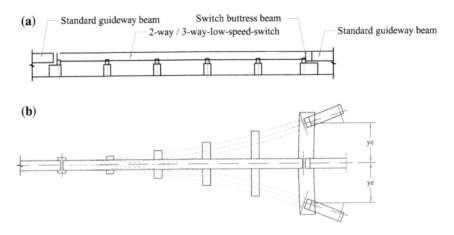

Fig. 5.87 Bending type switch of Transrapid

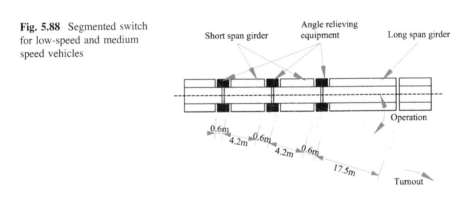

Fig. 5.88 Segmented switch for low-speed and medium speed vehicles

For low-speed vehicles, the segmented configuration in Fig. 5.88 is used [11]. It features a smaller curve radius, but it inevitably has discontinuity of the guideway, resulting in a sudden airgap signal or mechanical contact. In addition, the shorter section, made mainly of steel, may induce some vibration problems due to vehicle/guideway dynamic interaction.

References

1. Sinha PK (1987) Electromagnetic suspension dynamics and control. Peter Pergrinus Ltd., London
2. Lim J, Kim C-H, Han J-B, Han H-S (2015) Design of an electromagnet with low detent force and its control. J Electr Eng Technol 10(4):1668–1674
3. Ha C-W, Kim CH, Cho JM, Lim JW, Han HS (2015) Magnetic levitation control through the introduction of bogie pitch motion into a control law. J Korean Soc Railway 18(2):87–93

4. US Department of Transportation (2004) Chubu HSST Maglev system evaluation and adaptability for US urban Maglev. FTA-MD-26-7029-03.8
5. Zhao CF, Zhai WM (2002) Maglev vehicle/guideway vertical random response and ride quality. Veh Syst Dyn 38(3):185–210
6. Han HS, Yim BH, Lee NJ, Hur YC, Kim SS (2008) Effects of the guideway's vibrational characteristics on the dynamics of a Maglev vehicle. Veh Syst Dyn 47(3):309–324
7. Zhou Danfeng, Hansen Colin H, Li Jie, Chang Wensen (2010) Review of control vibration problems in EMS Maglev vehicles. Int J Acoust Vib 15(1):10–23
8. Zhai W, Zhao C, Cai C (2006) Dynamic simulation of the EMS Maglev vehicle-guideway-controller coupling system. Maglev 2006, Dresden, Germany
9. Han HS, Yim BH, Lee NJ, Kim YJ (2009) Prediction of ride quality of a Maglev vehicle using a full vehicle multi-body dynamic model. Veh Syst Dyn 47(10):1271–1286
10. Yim BH, Han HS, Lee JK, Kim SS (2009) Curving performance simulation of an EMS-type Maglev vehicle. Veh Syst Dyn 47(10):1287–1304
11. Park Doh Young, Shin Byung Chun, Han Hyungsuk (2009) Korea's Urban Maglev program. Proc IEEE 97(11):1886–1891

Chapter 6
Railway Applications

6.1 Introduction

Maglev technology has evolved most significantly in the area of railway passenger transport because of its emergence as an alternative to wheel-on-rail systems. Its development resulted in the construction of a number of systems currently in service and under construction. Since 2004, the Transrapid, the first maglev-based commercial rail system in the world, has been operating in Shanghai, China at a maximum operating speed of 430 km/h without any safety problems. Through its 10 years of successful operation, it has proven that the electromagnetic levitation system is technically mature and reliable. For urban transit, the Linimo has been in service in Nagoya, in Japan, carrying 20,000 passengers a day since 2005. During its 10 years of service, it has showcased the main benefits of urban maglev trains, such as low environmental impact, superior ride comfort and low maintenance cost. In Korea, the urban transit ECOBEE has been undergoing test runs in advance of its eventual opening, which will likely be in 2016. Two urban maglev lines are now under construction in China, which is aiming at opening them in 2016. It can be said that the maglev train will be a new railway transportation mode for the future. On the other hand, with the improvements in speed and ride comfort that have been achieved by wheeled trains, the distinction between maglev and wheeled trains is shrinking. As such, maglev systems must be significantly innovated to retain their inherent superiority. In this chapter, the applications of maglev to railways are outlined. Though there is a wide range of systems, only commercial and operational systems are dealt with, as the aim is to highlight the practical aspects. In addition, systems that have a long history are presented briefly because of their extensive coverage in the literature or on the Internet.

© Springer Science+Business Media Dordrecht 2016
H.-S. Han and D.-S. Kim, *Magnetic Levitation*,
Springer Tracts on Transportation and Traffic 13,
DOI 10.1007/978-94-017-7524-3_6

6.2 Transrapid

6.2.1 Introduction

The well-known German Transrapid was commercialized in 2004 in Shanghai, approximately 40 years after research for its development began. Its main technical feature is the use of electromagnets and LSM for levitation and propulsion, respectively. Its state-of-the-art levitation control system maintains the clearance between the lift magnet and the guideway at 10 mm. Thanks to the active position control, the guideway following characteristic, in addition to good ride comfort, allows for a smaller radius of curve, which gives its operators flexibility in formulating routes. The status of Transrapid can be summarized into two aspects. First, through its 10 years of successful operation, it has proved that the electromagnetically levitated train can be a mode of railway transportation. Second, to reduce the initial capital cost and maintenance costs to the level of the conventional systems, the adaptation of the electrical components for electric cars and industrial systems, i.e. the development of standard components, is being proposed. A detailed account of the Transrapid is not given here, as thanks to its nearly 40-year development history, the related technical information is available in the public domain, including on the Internet. As such, the contents in this section emphasize the three aspects mentioned above.

6.2.2 Background

Transrapid can be described as the world's most advanced electromagnetic system. It uses electromagnets both for suspension and guidance, and the iron-core long-stator LSM for propulsion, which enables it to reach speeds of more than 500 km/h.

- **Levitation/Guidance**: Electronically controlled support magnets located on both sides along the entire length of the vehicle pull the vehicle up to the ferromagnetic stator packs mounted to the underside of the guideway (Figs. 6.1 and 6.2). Guidance magnets located on both sides along the entire length of the vehicle keep the vehicle laterally on the track. Electronic systems guarantee that the clearance remains constant (nominally 10 mm). Incredibly, the Transrapid requires less power to levitate than is used by its air conditioning equipment. The levitation system is supplied from on-board power supply (batteries, inductive power supply or linear generators) and thus is independent of the propulsion system.
- **Propulsion**: The synchronous long-stator linear motor of the Transrapid maglev system is used both for propulsion and braking. It functions like a rotating electric motor whose stator is cut open and stretched along under the guideway. Inside the motor windings, alternating current generates a magnetic traveling

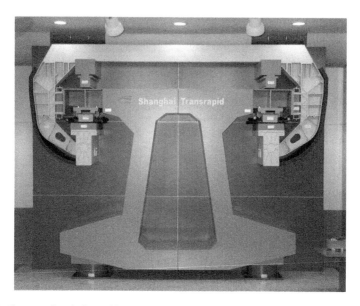

Fig. 6.1 Cross-sectional view of Transrapid

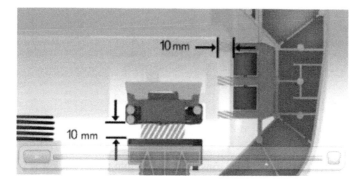

Fig. 6.2 Levitation, guidance and propulsion concepts of Transrapid using electromagnets [1]

field which moves the vehicle without contact (Fig. 6.3). The support magnets in the vehicle function as the excitation portion (rotor). The propulsion system in the guideway is activated only in the section where the vehicle actually runs (Fig. 6.4). The speed can be continuously regulated by varying the frequency of the alternating current. If the direction of the traveling field is reversed, the motor becomes a generator which stops the vehicle without any contact. The braking energy can be re-used and fed back into the electrical network.

- **Power supply**: While travelling, the on-board batteries are recharged by linear generators integrated into the support magnets. At low speeds, since the generated power is not enough, the induction type power supply is provided, as will be described later.

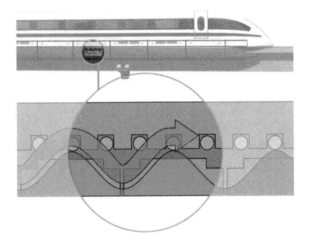

Fig. 6.3 Propulsion concept of Transrapid using a long-stator LSM [1]

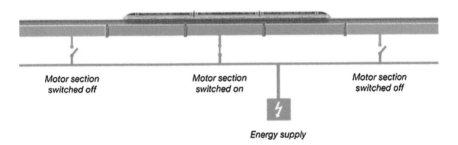

Fig. 6.4 Activation of the guideway for propulsion [1]

- **Guideway**: The Transrapid runs over a double-track guideway. It can be mounted either at grade, or elevated on slim columns, and consists of individual steel or concrete beams up to 62 m in length (Fig. 6.5).
- **Switch**: The Transrapid maglev system changes tracks using steel bendable switches (Fig. 6.6). These consist of continuous steel box beams with lengths between 78 and 148 m (256–486 ft), which are elastically bent by means of electromagnetic setting drives and securely locked in their end positions. In the straight position, the vehicle can cross the switch without speed restrictions; in the turnout position, the speed is limited to 200 km/h (125 mph) (high speed switch) or 100 km/h (62 mph) (low speed switch).
- **Operational control system**: The operation control system controls the operation and guarantees the safety of the Transrapid system. It safeguards vehicle movements, the position of the switches, and all other safety and operational

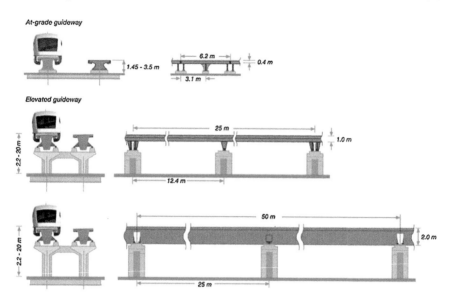

Fig. 6.5   Guideway designs of Transrapid [1]

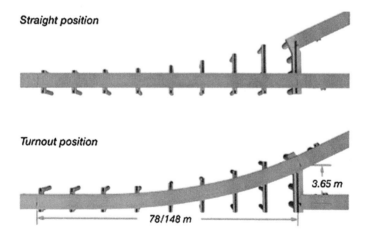

Fig. 6.6   Configuration of the bending type switch for Transrapid [1]

functions. The vehicle's location on the track is identified using an on-board system that detects digitally encoded location flags on the guideway. A radio transmission system is used for communication between the central control center and the vehicle (Fig. 6.7).

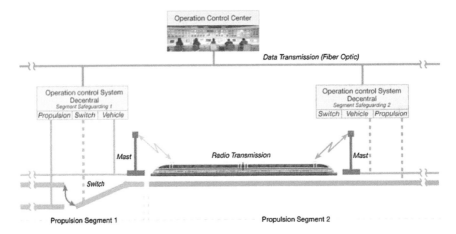

Fig. 6.7 Scheme for operation control system [1]

6.2.3 Operational Achievements

As of the spring of 2014, the Shanghai Maglev Transrapid Line had successfully achieved 10 years of continuous commercial operation, despite extreme weather conditions. Over its life so far, it has carried approximately 40 million passengers and traveled a running distance of around 12 million km. This maglev line has an impressive punctuality rate of 99.93 %, and has achieved significantly high speeds and high transport volume. Although it was designed as a demonstration line, the line makes an important contribution to Shanghai's public transport network.

6.2.4 Technical Features

Before describing the general specifications and features of Transrapid, the recent status of technological development achieved over its 10 years of operation is outlined here [2]. For maglev trains now, the challenging question is whether maglev can ever achieve a clearly lower cost level than a conventional rail system, with better performance. To overcome this challenge, one of the recently proposed development strategies is to design the Transrapid system in such a way that existing large series components can be used, rather than designing components according to Transrapid system's need. As a result, maglev will benefit from the significant efforts of the automotive industry to establish large series of new generations of batteries that offer quality, high power and energy density at reasonable price. For power supply, the challenge is to adapt the on-board power supply in such a way that automotive battery containers can be used for maglev without any alteration, resulting in the removal of the contact or non-contact power supply mechanism. In another example, the current centralized magnet driver with a capacity of 440 V and 80 A for 10 magnets can be decentralized into a smaller unit for a single magnet using well-developed automotive components. This concept is called "mechatronic magnet poles," in which each magnet pole has its own electronic control (driver). Adapting the standardized components could reduce the cost and volume dramatically. This also allows for the removal of the electrical brake, an eddy current type brake, for emergencies, using the guidance magnet for braking instead. For the propulsion system, the current configuration is characterized by inverters installed in substations, and power distribution via cable and switch units along the track every 1–2 km. The inverter units are far beyond standard industrial types, and demand costly special equipment. But today, with the development of power electronics in electric cars and industrial applications, the idea of decentralized inverters can be applied. That is, the application of large series products for industrial applications with ratings of 1 kV and 200–500 kW per device may become feasible, and a modular concept can be realized. Based on this modular concept, the inverters could be integrated into the guideway beams. Guideway beams containing inverters could be manufactured with complete propulsion equipment as one autonomous mechatronic unit, which would have the effect of reducing total construction costs. Eventually, it is likely that the initial capital costs of a maglev system could be lowered to the level of conventional systems.

A brief outline of some of the main design features of the latest Transrapid 09 is relevant here (Fig. 6.8 and Table 6.1).

Fig. 6.8 Latest Transrapid TR09 [3]

Table 6.1 TR09 main specifications [3]

Item	Value
Number of sections	3
Total length	75.8 m
Vehicle width	3.7 m
Vehicle height	4.25 m 3.35 m (from guideway gradient)
Inner width of carriage body	3.43 m
Inner height of carriage body	2.1 m 2.05 m (entrance door area)
Empty weight	169.6 tons
Full loaded weight	210 tons
Design speed	505 km/h
Passenger capacity	449 persons
Design pressure	±5500 Pa
Sealing time constant	$\tau > 20$ s

- Increase the payload to meet peak hour demand by allowing standing passengers.
- Decrease interior noise level through the application of several measures, which are sound isolation of outer skin of carriage body, optimization of acoustics, and reduction of sound sources.
- Increase of the inner width and space of entrance doors for airport shuttle requirements.

Fig. 6.9 Inductive power collection system for low speeds [3]

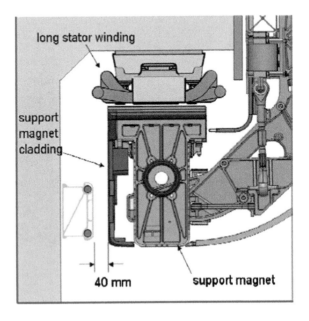

- Realization of the exterior and interior appearance in close collaboration with Deutsche Bahn.
- Safety assessment by EN 5012× and Maglev design rule from German Federal Authority.
- IPS (Inductive power supply) is employed at lower speeds (Fig. 6.9).

6.3   Linimo

6.3.1   Introduction

The Japanese urban transit Linimo using electromagnets and LIM has been successfully operated over the Tobu Kyuryo Line in Nagoya since 2005 (Fig. 6.10). Carrying 20,000 passengers a day, the operator is earning profits over the operating costs. In the mid-1970s, the development program at JAL for Linimo was prompted by the need for a new system for the 66 km route between the new international airport at Narita and Tokyo. This long history ultimately resulted in the Linimo and the Tobu Kyuryo Line. The system configuration of Linimo is best illustrated by HSST-100L, one of the prototypes of Linimo, in Fig. 6.11. The guideway consists of a U-shaped rail for levitation, a reaction plate for LIM, sleeper, girder, power rail, and signaling cable. The vehicle is composed of 5 bogies, with each bogie having a levitation/propulsion module, air springs, and associated components. Only the main features are highlighted.

Fig. 6.10 Linimo operating on Tobu Kyuryo Line in Nagoya, Japan [4]

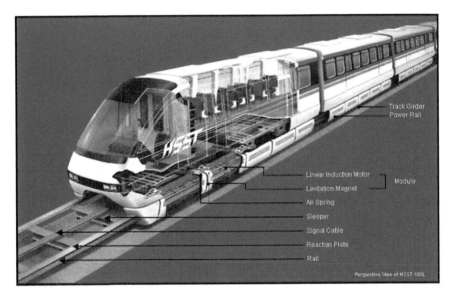

Fig. 6.11 General configuration of CHSST-100L (prototype of Linimo) [5]

6.3.2 *Vehicle*

The technical specifications of Linimo are given in Table 6.2 and its devices are presented in Table 6.3.

- Levitation and guidance: The suspension and guidance forces are provided by a typical U-core electromagnet, which is described in Chap. 5 (Fig. 6.12) [5]. The details of design and levitation control principles are given in the previous Chap. 5 and the subsequent Sect. 6.3. The nominal and landed airgaps are 8 and 14 mm. The input voltage to the magnet driver is 275 VDC. The capacity of the levitation magnet is 1040 kgf/m. The levitation control algorithm is given in Fig. 6.13, where the acceleration, airgap, and airgap velocity are fed back [8]. That is, the control voltage is derived by multiplying the three control gains by the three signals.

   Propulsion: The thrust needed for propulsion is generated by 10 LIM per car. The technical specifications of the LIM are listed in Table 6.4. The rated thrust per car is 39.8 kW/car, and the connection is 5S2P. One VVVF with IGBT and PWM is used per car. Although the current control is essentially achieved by the thrust command calculated using ATO notch signal or manual notch signals, the thrust command is compensated by the vehicle weight data, which is picked up by air suspension pressure transducers. In the range of high speed, the thrust is limited due to the limitations of the output voltage of the inverter. The output frequency is also an important factor determining the speed of the vehicle. Non-contact speed detection devices are used as a source of speed data for the inverter system. There is

Table 6.2 Linimo's technical specifications [6, 7]

Size	Train length		43.3 m
	Car length: head section		14.0 m
	Car length: middle section		13.5 m
	Width		2.6 m
	Height		3.445 m (above the skid surface of the rail)
Vehicle performance	Maximum operational speed		100 km/h
	Acceleration	Maximum	4.0 km/h/s (w/passenger load compensation)
	Deceleration	Full service break	4.0 km/h/s (w/passenger load compensation)
		Emergency break	4.5 km/h/s (w/passenger load compensation)
		Back-up break	4.8 km/h/s (empty)
Maximum gradient			60 ‰
Minimum curve radius			75 m
Minimum passing horizontal curve radius			1500 m
Empty weight			17.0 tons/car
Maximum design weight			28.0 tons/car (fully loaded)

Table 6.3 Devices on Linimo [6, 7]

Classification	Device	Remarks
Levitation control	MDU (magnet drive unit)	1/module (10/car)
Propulsion	VVVF inverter control	Propulsion and electric brake
Mechanical brake	Hydraulic brakes	Operates under 5 km/h
Electric converter	PSU (power supply unit): H-INV, L-INV, AC-INV	DC 275 V, DC 100 V, AC 100 V
Train operation	• TD (train detector) • ATC, ATO • VEL (velocity) • Platform control • Wireless system • IR (inductive radio)	Information between train and ground
Control monitor	TIMS	Total train control

a predetermined slip frequency that represents the difference between the inverter output frequency and synchronous frequency, which is proportional to the vehicle speed. The frequency applied to the LIM is controlled both in acceleration and deceleration ranges (Fig. 6.14 and Table 6.5).

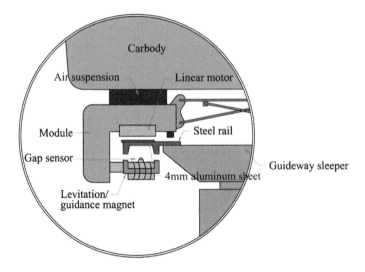

Fig. 6.12 CHSST maglev rail and module cross-section

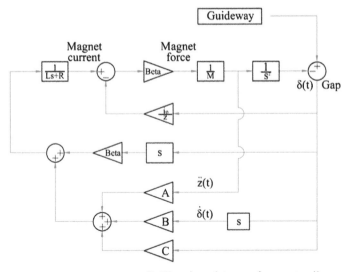

R: Electric resistance of magnet coil
L: Total Inductance of coil
Z: Target air gap of magnet
i0: Typical current of magnet coil
Beta: Magnet gain
M: Mass of bogie
C: Proportional feedback gain of gap
B: Differential feedback gain of gap
A: Acceleration feedback gain

Fig. 6.13 Block diagram of the controller of Linimo

Table 6.4 LIM specification of Linimo [6]

Thrust (nominal)	3000 N/LIM
Phases/poles	3 phases/8 poles
Material of coil	Aluminum
Current (Max)	380 A
Length	1.8 m
Width	0.6 m
Thickness	0.08 m
Secondary conductor (track)	Thickness: 4 mm Width: 240 mm Material: aluminum

Fig. 6.14 Photo of LIM for Linimo [7]

Table 6.5 Major design specifications of the VVVF inverter [6]

Trolley rail voltage	1500 V DC
Max. output voltage	1100 V AC
Max. capacity	1450 kVA/2 car unit
Frequency range	0–90 Hz
Type of control	Voltage control type, Pulse width modulation (PWM)
Cooling type	Forced air-cooling

- **Braking**: The brake works through a combination of three braking modes. The first braking mode is electric service braking using the LIM, and the second is the hydraulically controlled mechanical brakes [6]. The third braking mode is on the landing skids, in which the vehicle is de-levitated. The electric service brakes themselves operate in one of two modes depending upon vehicle speed. The first mode is the regenerative mode, which normally operates at higher vehicle speeds, and the second mode is the dynamic brake mode, which normally operates at lower speeds. The mechanical brakes are hydraulically controlled caliper brakes in which redundant and independently controlled hydraulic power sources are located on each vehicle. The calipers grip the outer rail flange. The mechanical brakes are also used as the parking brake. During high-speed operation, the normal braking sequence is to first apply electric brakes. When a lower speed is reached, the mechanical brakes are incorporated, with braking becoming fully mechanical at very low speeds.

Bogie: Linimo has 5 bogies, as shown in Fig. 6.15 [5, 7]. One of the outstanding bogie designs is the optimized and wired steering mechanism (Fig. 6.16) [7]. This was optimized to attain both ideal curve negotiation capability and cross-wind resistance. Two considerations are incorporated into the bogie mechanism. First, allowing for side forces due to cross-wind, the lateral displacements between rail and magnets must be equally distributed through the length of the vehicle over the straight track. Second, during curve negotiation, the magnets should follow the rail

Fig. 6.15 General view of Linimo bogie [7]

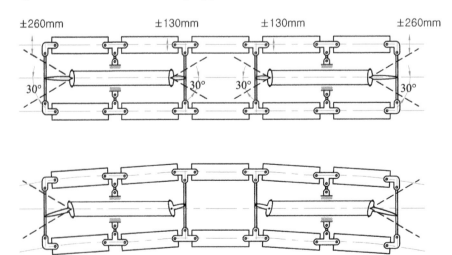

Fig. 6.16 Bogie steering mechanism of Linimo

profile with various radii of curves. Linimo consists of 5 bogies interconnected mechanically to satisfy the above two objectives. The ratio of lateral movement of the leading bogie to the central one is adequatley chosen. While running over the straight section under side wind forces, through the mechanical connections and its movement ratio, the lateral displacements of all the magnets are expected to be equal, resulting in even guidance force distributions. During curve negotiation, the capability of following the guideway profile with the various radii of curves is much more important for an urban maglev vehicle. Some assumptions are made in order to select the optimum movement ratio described earlier. The adequate ratio must be chosen to meet the guideway profile following capability based on the above assumptions. Fortunately, the ratio derived from curve negotiation is the same as that derived from running over a straight section under side wind forces.

6.3.3 Guideway

The baseline guideway in both the test track and in planned applications is elevated, and consists of a simple box girder for each travel direction topped with transverse steel sleepers, which in turn support the maglev rails described above. Other than tunnel, the elevated configuration is most preferred for urban/suburban infrastructure compatibility (Fig. 6.17). Two-way elevated guideways consist of two parallel guideway beams, supported on traditional cross-beams and pylons/footings, designed for local conditions and long-term stability. All services, such as power

Fig. 6.17 General configuration of guideway of Linimo [5]

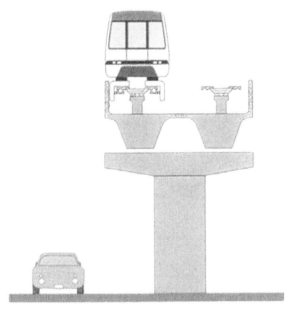

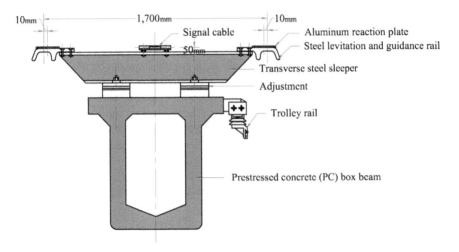

Fig. 6.18 Configuration of guideway [5]

transmission, signal and communication, etc. are located on the guideway. Rights of way of existing major streets can thus be utilized. The arrangement of the basic guideway beam, rails, sleepers, and support is shown in Fig. 6.18 [5]. The steel rail section, specially designed, provides both the levitating two-pole lower section and the upper LIM surface, covered with aluminum (insulated from steel), with the outer vertical flange also used for mechanical brakes Fig. 6.19 [7]. Guideway rail alignment can be done via adjustments in the seating of the sleepers on the beams. The power feeder line is placed on the sides of the main beam (Fig. 6.20).

6.3.4 Deployment

Linimo may be evaluated as an optimized system for urban passenger transportation thanks to its more than 30-year evolution. Tobu Kyuryo Line is shown in Fig. 6.21.

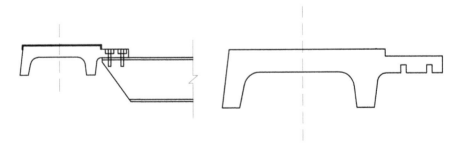

Fig. 6.19 Dimensions of U-shaped rail for levitation [7]

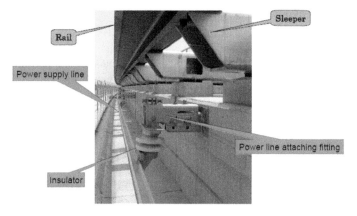

Fig. 6.20 Power rail assembly [7]

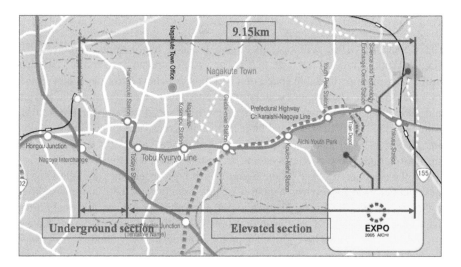

Fig. 6.21 Tobu Kyuryo Line [9]

The effectiveness of the deployment of Linimo has already been described earlier. Its inherent low noise and vibration as well as its lowest maintenance cost have been proven through real operations. While there have been several attempts to deploy it in other countries, due to the global economic crisis in 2009 and a number of other factors, these were unfortunately discontinued.

Table 6.6 Tobu Kyuryo Line specifications [9]

Items	Description
Route	Fujigaoka to Yakusa: 9.2 km Service length: 8.9 km, 9-stations
System	EMS (HSST-100) Max. speed: 100 km/h Trip time: approx. 15 min
Track	Double track, elevated (partially tunnel 1.3 km)
Put into service	2005
Construction cost	Total construction cost: approx. 100 bil JPY Infra part: approx. 60 bil JPY Non-infra part: approx. 40 bil JPY

6.4 ECOBEE

6.4.1 Introduction

The Korean ECOBEE will operate on the Incheon International Airport Urban Maglev Demonstration Line (Fig. 6.22). The system, operating over a 6.1 km track with a maximum permissible speed of 110 km/h, is believed to offer a new option for urban transit. The maglev development program in Korea was opened in the mid-1980s, prompted by the need for a new high-speed railway for the 420 km route between the first and second largest cities in Korea, Seoul and Busan. At that time, the high-speed maglev system was one of the candidates for the project. For this reason, high-speed maglev systems began to attract interest in Korea. However, due to the selection of TGV as the Seoul-Busan high-speed line, and allowing for the potential need for environmentally-friendly urban transit between many growing cities nationwide given Korea's rapid economic growth at the time, urban maglev work began to be studied in late 1989 with the support of the Korean government. Based on the significant progress achieved over about two decades, the realization program named Urban Maglev Program was started from late 2006 [10]. The Urban Mgalev Program is funded by the Korean Ministry of Land, Infrastructure and Transport (MOLIT) with assistance from the Korean Ministry of Trade, Industry and Energy (MOTIE) and the Ministry of Science, ICT and Future Panning (MSIP), and under the supervision of the Korea Agency for Infrastructure Technology Advancement (KAIA). The total program budget is expected to be 450 million USD, and includes contributions from the private sector, Incheon International Airport Corporation and City of Incheon. The Program is composed of three Core Projects: Systems Engineering, Vehicle Development and Demonstration Line Construction. The Core Projects are managed by KIMM, Hyundai-Rotem and Korea Rail Network Authority, respectively. The technical features and deployment of ECOBEE are briefly presented in this section.

Fig. 6.22 ECOBEE on the 6.1 km long Incheon International Airport Urban Maglev Demonstration Line

6.4.2 Background

ECOBEE is a typical electromagnetic attraction system, as described in Chap. 5 and Sect. 6.3, with U-shaped magnets and LIMs for levitation and propulsion (Fig. 6.23). The guidance forces are provided by the lift magnets, which are proportional to lateral displacement. The thrust for acceleration and deceleration is achieved by the LIMs, with the on-board primary windings and the secondary reaction plate in the guideway. The mechanical brakes are also used with the LIMs. The operation scheme is almost the same as that of Linimo, which is described in Sect. 6.3. In addition, the vehicle has landing skids for emergency landing and landing rollers for rescue operation by other vehicles.

- **Guideway**: The standard guideway is a type of elevated guideway with a double track. The 105 m long main beam in each direction is supported at regular intervals of 35 m with pylons/footing. Then the two main beams are connected with cross-beams, resulting in the increased stiffness and mass of the guideway. On the cross-beams, a walkway is installed to offer a way for passenger evacuation and maintenance works. The sleepers are placed on the beams and in turn support the U-shaped rail for magnetic suspension. The power feeder lines are installed on both sides of the main beams. This configuration could be constructed over the existing roads with less acquisition of rights of way.
- **Vehicle**: The design specifications of ECOBEE are summarized in Table 6.7. The maximum design speed is 110 km/h and the speed was attained by the

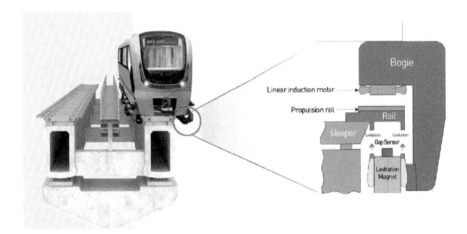

Fig. 6.23 ECOBEE levitation module and rail cross-section

Table 6.7 Technical specifications of ECOBEE

Items	Specifications
Formation	– 2 cars
Track conditions	– Gauge: 1850 mm – Max. gradient: 7 ‰ – Min. curve radius: 50 mR – Power supply: 1500 VDC (3rd rail)
Train performances	– Max. design speed: 110 km/h – Max. operating speed: 80–100 km/h – Max. acceleration: 4.0 km/h/s – Max. deceleration: 4.0 km/h/s (4.5 km/h in emergency) – Cabin noise level: 70 dB(A) – Ride index: below 2.0 of UIC – Nominal airgap: 8 mm
Number of bogies	– 4 per car
Suspension	– Electromagnetic
Propulsion	– LIM + VVVF
Braking system	– Blending of regenerative and pneumatic brake
Vehicle dimensions	– 12(L) × 2.7(W) × 3.475(H) m
Passenger capacity	– 115 passengers/car (including 22 seats)
Weight	– Empty: 19 tons/car – Full: 26.5 tons/car
Operation	– ATO/driverless

running tests. Some detailed accounts of each technical item are briefly
described in the following sections.

- **Potential applications**: ECOBEE has potential applications in urban or suburban in which environmental concerns have a greater priority.

6.4.3   Levitation and Guidance

Having described the most of the principles of levitation and propulsion for ECOBEE in Chap. 5, only some of the main features are introduced here.

- **Lift magnet**: The electromagnet for ECOBEE, shown in Fig. 6.24, is composed of two poles, 4 yokes and 8 windings. The U-shaped rail facing the magnet has the profile shown in Fig. 6.25. The technical specifications of the magnet and its associated driver are listed in Table 6.8. Long-pole design is employed to reduce the magnetic drag forces. The lift and guidance force characteristics were measured with a 1/4 magnet, as shown in Figs. 6.26 and 6.27, respectively. The nominal current for empty load is measured to be about 30 A, while it is 35 A for full load, with nominal airgap of 8 mm.
- **Levitation control**: The levitation control is based on the 5 states feedback scheme described in Chap. 5. The measurement of airgap and acceleration is available through an external sensor module containing two gap sensors, and an accelerometer is placed in between two poles (Fig. 6.28). Two inductive gap sensors are used to output a smoother signal from two measured signals by switching them, avoiding discontinuity in measured signals at rail joints.

Fig. 6.24 General view of the lift electromagnet of ECOBEE

Fig. 6.25 Rail profile for levitation

Table 6.8 Specifications of ECOBEE's support magnet

Items	Values
Pole length	2600 mm
Pole width	32 mm
Clearance between poles	154 mm
Pole height	128 mm
Number of yokes per pole	4
Number of coils per pole	2
Coil turns	193
Magnet height	191 mm
Rated lift force	33 kN
Nominal current	27 A
Nominal airgap	8 mm
Magnet driver input voltage	DC 350 V
Magnet driver	2-quadrant chopper operating at 5 kHz
Gap sensor	Inductive proximity type: 3–19 mm
Accelerometer	Semiconductor type: +5 to −6 g
Control algorithm	State feedback

Fig. 6.26 Lift force-current characteristics at different airgaps

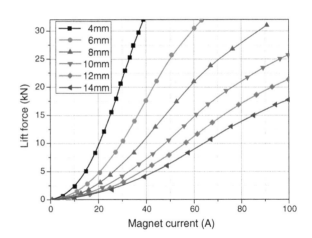

6.4.4 Propulsion System

The design requirements for propulsion are listed in Table 6.9.

The drag resistance needs to be identified before designing the LIM. Composite drag D per car is composed of four terms.

$$D = (D_c + D_m + D_a + D_g)[N] \qquad (6.1)$$

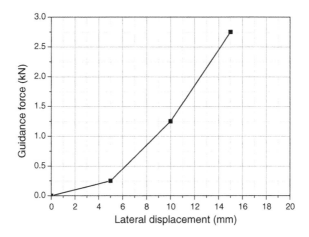

Fig. 6.27 Guidance force-lateral displacement with nominal point (8 mm, 30 A)

Fig. 6.28 Sensor module consisting of two gap sensors and an accelerometer inside

where,

Power collector drag $D_c = 41.68 \times n$

Magnetic drag $D_m = 3.354 \times n \times W \times V \, (for \; 0 \leq V \leq 5.6 \text{ m/s}) \; and$

$$18.22 + (0.074 \times V)W \times n \, (for \; V \geq 5.6 \text{ m/s})$$

Aerodynamic drag $D_a = (1.6522 + 0.572 \times n)V^2$

Gradient drag $D_g = n(W \times g) \sin \theta$

- n number of cars
- g gravitational acceleration
- V vehicle speed
- W weight of car

The drag forces calculated by Eq. (6.1) are given in Fig. 6.29. The LIM was designed to meet the propulsion requirements considering the drag forces calculated and listed in Table 6.10. The constant slip frequency control algorithm is employed in the LIM. The thrust required can be controlled by adjusting the LIM currents with a constant slip frequency of LIM, and the normal forces can be limited within a certain range. The configuration is shown in Fig. 6.30. The required slip frequency is derived from the thrust and normal forces—slip frequency characteristics given in Fig. 6.31. The resulting thrust performance is presented in Fig. 6.32, which shows the capability of 110 km/h at zero-gradient and 62 km/h at 7 ‰ slope. In addition, an additional thrust force needed to draw a disabled train was included in the resulting thrust.

Table 6.9 Propulsion requirements

Requirements and characteristics	Values
Vehicle weight (full load)(W)	26.5 tons/car
Maximum speed	110 km/h
Maximum initial acceleration	4 km/h/s
Maximum gradient	7 ‰
Motor current	Max. 380 A
Inverter output voltage	1100 V
Required max. tractive effort	60,400 N/train

Fig. 6.29 Drag force characteristics (train)

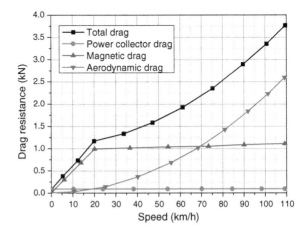

Table 6.10 LIM design specifications

Item	Value
Length	1785 mm
Number of poles	8
Number of slots	53
Turns per coil	5
Tooth width	12.5 mm
Slot width	21 mm
Thickness of reaction plate	5 mm
Height of reaction plate overhang	20 mm
Airgap	11 mm
Slip frequency (acceleration)	12.5 Hz
Slip frequency (deceleration)	13.7 Hz

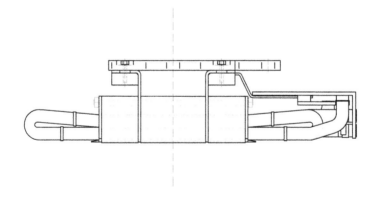

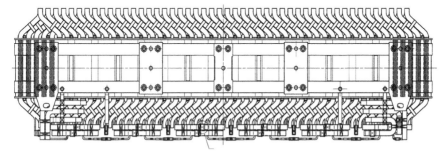

Fig. 6.30 LIM configuration

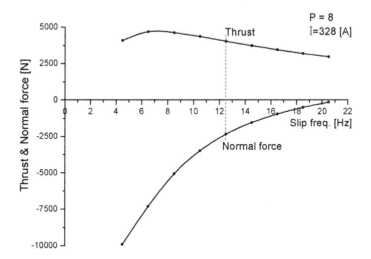

Fig. 6.31 Thrust and normal forces—slip frequency characteristics

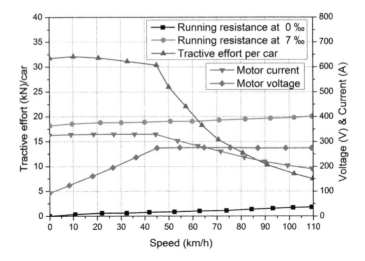

Fig. 6.32 Thrust capability

6.4.5   Guideway

The basic guideway structure for ECOBEE is illustrated in Fig. 6.33. Some details of this are described in the previous section. Figure 6.34 shows the kinematic clearances, of which the centerline distance is 4.5 m. Though the ECOBEE basic guideway is an elevated type, it can be also applied in a tunnel or at-grade, as shown in Fig. 6.35.

Fig. 6.33 Cross section of the ECOBEE basic guideway

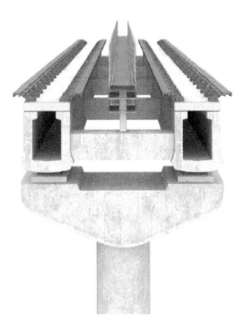

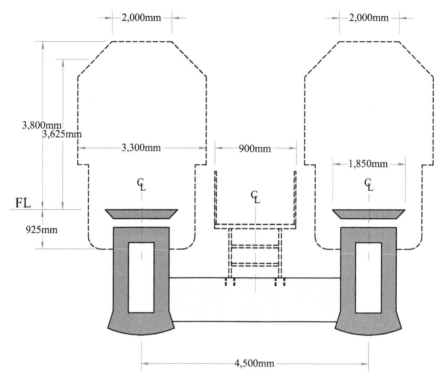

Fig. 6.34 Construction gauge

The track is composed of levitation rails, reaction plates, sleepers, plinths, fixtures and girder. The track gauge is 1850 mm between the rail centerlines. Rail alignments can be made through adjustment of the plinth. The sleeper spacing is 1.15 m considering the length of the levitation module and its strength. The profiles of the levitation rail and reaction plate are given in Figs. 6.36, 6.37 and 6.38, respectively, and its basic lengths are 10 and 12.5 m. Reaction plate lengths are 5 and 6.25 m. The guideway's geometric tolerances are listed in Table 6.11.

The ECOBEE guideway can be characterized as follows.

- **Mass**: Due the nature of the electromagnetic attraction system, the smaller vehicle/guideway mass ratio has a positive effect on the suspension stability. In contrast, because decreasing the ratio makes the guideway heavier, the design of the guideway is based on a compromise between the guideway construction costs and the stability and ride comfort. For ECOBEE, the vehicle/guideway mass ratio was chosen to be basically lower than 0.6.
- **Deflection**: Allowing for ride comfort and the vehicle/guideway mass ratio, the deflection limit at the mid-span for the basic guideway is set to be L/2000 mm, and to be L/3000 m for relatively higher speeds.

Fig. 6.35 Guideway in tunnel

- **Main beam**: The 3 span continuous beam increased effective mass over a span and decreased the guideway deflection, resulting in an improvement in ride comfort.
- **Natural frequency**: The fundamental natural frequency of the basic guideway is 3.4 Hz, which is a sufficient distance from the electromagnet's natural frequency of around 10 Hz.
- **Compatibility**: The guideway structure must be in harmony with the surrounding buildings, minimizing aesthetic impact and construction cost.
- **Curves**: The Demonstration Line consists of 3 curves with the cants and transition lengths for each, as shown in Table 6.12.

6.4.6 Vehicle

The ECOBEE consists of 2 cars, and the dimensions are shown in Fig. 6.39 [11]. The main features of the ECOBEE can be summarized as follows.

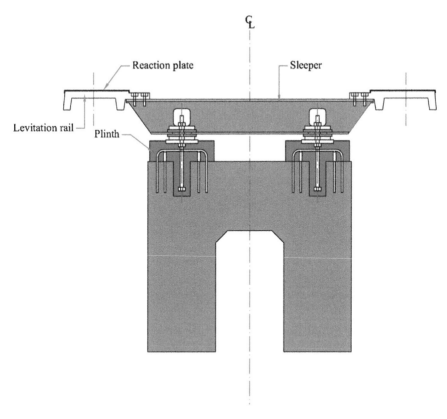

Fig. 6.36 Configuration of track

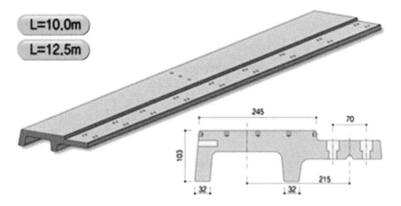

Fig. 6.37 Steel levitation and guidance rail

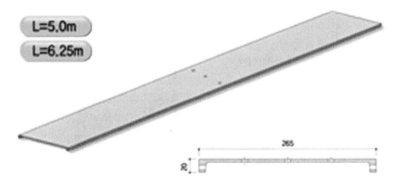

Fig. 6.38 Reaction plate with back iron for LIM

Table 6.11 Guideway tolerances

Irregularities	Tolerances
Deviation from alignment (vertical)	±3 mm (chord length = 10 m)
Deviation from alignment (lateral)	±3 mm (chord length = 10 m)
Track gauge	±3 mm
Rail joint alignment (vertical)	±1.0 mm
Rail joint alignment (lateral)	±0.5 mm
Level difference	±2 mm

Table 6.12 Curves on Incheon International Airport Maglev Demonstration Line

Curve radius (m)	Max. speed (km/h)	Cant	Transition length (m)
125/110		2.4°	56
150	35	2.7°	63
800	80	3.4°	126

- **Exterior design**: The design principle of the train's exterior was chosen to highlight its shape. It is inspired by the curve found in Korean traditional celadon, and incorporates a honeycomb pattern (Fig. 6.40b).
- **Interior design**: Interior design and seat arrangement are customized for airport application, and thus considers that passengers are bringing their baggage (Fig. 6.40b).
- **Misted window**: Window shading function protects the privacy of urban residents (Fig. 6.41).
- **Backup battery**: When the main power is lost, the backup battery is able to provide levitation power for approximately 30 s. The 30 s is necessary for a safe landing.
- **Light weight carbody**: Single skin aluminum is used to achieve a reduction in carbody weight.

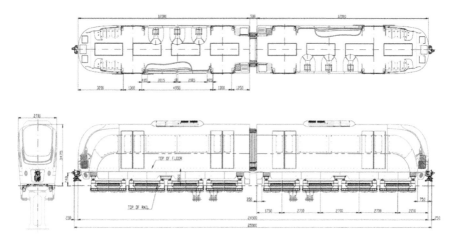

Fig. 6.39 Dimensions of ECOBEE

Fig. 6.40 Exterior and
interior design of ECOBEE

Fig. 6.41 Misted window

- **LED**: All the lights both inside and outside the vehicle use LED, reducing energy consumption.
- **Bogie**: The ECOBEE has 4 bogies, each having two levitation modules and 4 air springs. The air spring, a secondary suspension element, provides vertical and lateral stiffness and damping. No additional damper for secondary suspension is used because the damping of air springs is sufficient for ride comfort. For good track following capability of the magnet, the compliance between bogie frames is provided by inserting rubber bushing elements in joints and some linking mechanism. Another outstanding feature of bogies is the capacity to use a steering mechanism among them. There are two purposes of incorporating such a mechanism. The first one is to achieve the capability of negotiating lateral curves of 50 m, including 25 m on the switch. The second is to uniformly distribute side forces from crosswind to all the electromagnets for equal lateral displacements in all the magnets. To achieve these two purposes, hydraulic systems are introduced to make two levitation modules move relatively with a specified proportion. The concept is illustrated in Fig. 6.42, where a pair of levitation modules is connected through two hydraulic cylinders with an area ratio of 1:2.8. The ratio is optimized to smoothly follow a 50 m radius curve. When a crosswind is applied to the vehicles, the lateral displacements of the levitation modules are arranged as shown in Fig. 6.42b. It can be understood here that if more uniform lateral displacements of the magnets are required, then the number of bogies needs to be increased to 5, as in the Linimo. In addition, since it is difficult with the maintenance of the hydraulic system, it may need to be replaced as linkage or wires.

6.4.7   Braking

The ECOBEE has two kinds of braking modes, and it utilizes both electrical and mechanical brakes. These brakes operate either independently or in combination with each other. For electrical braking mode, the LIM is controlled to operate as an electric generator converting the vehicle's mechanical energy into electrical energy;

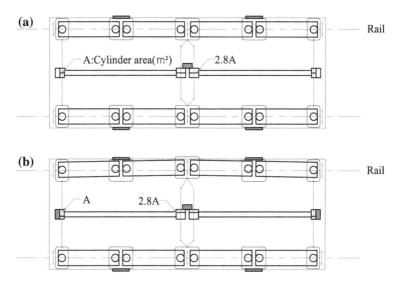

Fig. 6.42 Concept of bogie steering mechanism in ECOBEE: **a** undisplaced and **b** displaced positions

and furthermore, in dynamic mode, energy is supplied by the power supply for the plugging phase mode. The mechanical brake uses the calipers to grip the outer rail flange to obtain friction forces (Fig. 6.43). The materials for the mechanical brake pads should preferably have a high friction coefficient, and be easy to replace. The hydraulic power source is used for normal service braking. The mechanical brakes are also used as the parking brake. From high speed operation, the normal braking sequence is to first apply the electrical brakes. At a lower speed, the mechanical brakes are incorporated, and braking becomes fully mechanical at very low speeds. Operational experience has shown that the use of the mechanical brake should be minimized in order to reduce the replacement of the brake pads and particles. Finally, in the event of an emergency, the landing skids operate as a braking mode, both for stopping and landing. An emergency occurs when the vehicle is de-levitated, or when all other brakes have either failed or are not available. The material of the skid is chosen considering the friction and strength characteristics. For the ECOBEE, a sintered alloy is used in the landing skid assembly in Fig. 6.44.

6.4.8 Power Supply

ECOBEE operates with a 1500 VDC power supply connected to solid trolley rails (or power rail) (Fig. 6.45). The trolley rails are insulated above ground, and are exposed on each side of the guideway beam. Power collectors, i.e. pantographs, one for each of the insulated power rails, are placed on each vehicle (Fig. 6.46). The ECOBEE power collector configuration is relatively standard technology for

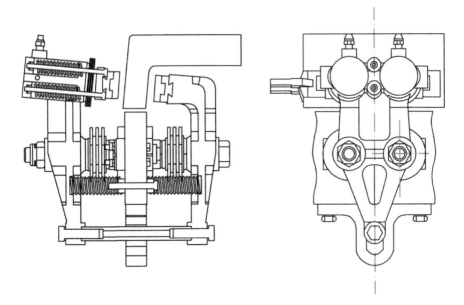

Fig. 6.43 Mechanical brake

Fig. 6.44 Landing skid made
of sintered alloy

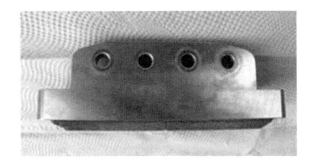

contact power collection. For continuous power collection, the current collecting
shoe in the collector is allowed to slightly move in normal direction to the power
rails by springs. Consequently, the collector follows the power rails with a constant
pressure until a certain speed is reached.

6.4.9 Automatic Train Operation

ECOBEE runs under ATO (Automatic Train Operation) driverless operation [11].
The signaling system has the capability of a minimum headway of 90 s. The ATO
system architecture is basically similar to that of conventional systems, except for

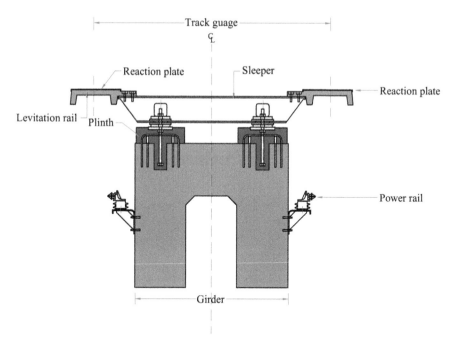

Fig. 6.45 Configuration and location of power rails

Fig. 6.46 Pantograph for current collecting

its contactless speed detection. The ATO requirements are given in Table 6.13. Speed detection and absolute distance detection use the pattern belt installed on the centerline of the guideway (Fig. 6.47a). Two speed ranges by two vehicle antennas are used to construct the detection logic in order to achieve a reasonable accuracy of detection. The speed and absolute distance are detected by counting and accumulating the two pulses from vehicle antennas. These two pieces of detected information provide the basis for ATO. The detailed account for the detection is well presented in Ref. [5].

Table 6.13 ATO
performance specifications

Items	Values
Train location detection accuracy	−5 to +10 m
Train speed detection accuracy	±3 km/h
Station stop accuracy	±0.3 m
Slip detection reference	less 2 m
Stop detection reference (for 2 s)	less 1–3 km/h

Fig. 6.47 Speed detection
system: **a** pattern belt on
guideway and **b** vehicle
antenna

(a)

(b)

6.4.10   Stray Magnetic Fields

One of the specific concerns in maglev vehicles is the stray magnetic fields produced by the magnet and electrical components. The magnetic field should be
limited for medical electronic device wearers and persons with cardiac pacemakers
and other implanted electronic devices. The magnetic field limits for ECOBEE are

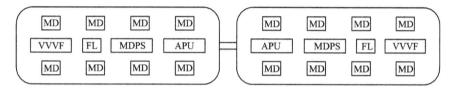

MD: Magnet Driver
APU: Auxiliary Power Unit
MDPS: Magnet Driver Power Supplier
FL: Filter Reactor
VVVF: Variable Voltage Variable Frequency

Fig. 6.48 Major components of ECOBEE

based on ICNIRP standards (1997) and the test method of EN 50500. According to ICNIRP standards, the static (DC) magnetic field below 1 Hz should be less than 40 mT, and over 1 Hz AC for time-varying magnetic field limits depending on frequency. The measurement points inside the vehicle are 30, 90, and 150 cm above the floor, higher than all the major electrical components (Fig. 6.48). For the DC magnetic field strength, the maximum value was measured to be 0.2 mT, 5 % of the guideline, at a height of 30 cm above the filter reactor for the VVVF inverter at 85 km/h. For AC magnetic fields, the VVVF inverter appeared to have the highest strength, as shown in Fig. 6.49, which indicates that the measured strengths over the frequencies are all below the guideline. It can be noted here that inside the vehicle, the highest strength stray magnetic fields are not directly related to the electromagnets. The reason is that the magnetic fields produced by the magnet are distributed in the vicinity of the airgap between the magnet and the ferromagnetic rail, as shown in Fig. 6.50.

6.4.11   Performance Tests

Vehicle performance tests are mandatory in Korea for safety. The planned performance test items are categorized into components function test, assembled vehicle performance verification tests on a short test track, and running tests over the main line including 5000 km reliability tests. In total, the tests cover 52 items, and all items were satisfied. The 52 test items are listed in Table 6.14.

- **Running performance**: Some test results related to running tests among the above 52 items are summarized in Table 6.15. Maximum operating speed for running performance tests was 80 km/h. The maximum design speed of 110 km/h was attained (Fig. 6.51).
- **Rescue operation test**: If a train is disabled, it must be rescued by another train. To verify this rescue operation capability, it was confirmed that the disabled train could be drawn by another train on a 4 per mil, grade (Fig. 6.52).

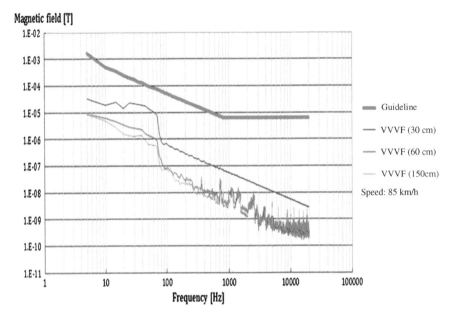

Fig. 6.49 AC time-varying magnetic field strengths measured

Fig. 6.50 Magnetic flux density distribution in the vicinity of an electromagnet

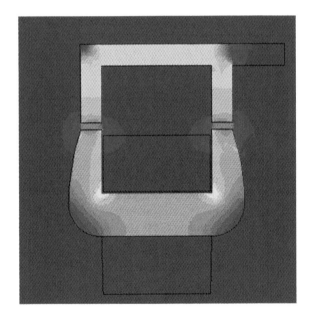

Table 6.14 ECOBEE Mandatory test items

Classification	Items	
Component	Structure load, bogie	
	Propulsion control equipment	Propulsion control inverter, traction motor, propulsion control system assembly
	Levitation equipment	Electromagnet, magnet driver Gap sensor
	Auxiliary power supply	APU-100, APU-330, MDPS
	Signaling equipment	On-board rack, ATO DU, ATO data processor, ATP/TP antenna, ATO local antenna, speed detection antenna, ATO data antenna
	Train control and monitoring system	CCU, TCMS DU, system (overall)
Finished rolling stock test	External structure	External structure, car body and bogie dimensions
	Measurement	Weight, vehicle kinematics, curving, wiring, insulation resistance, voltage endurance
	Car body leakage, car body lifting, propulsion control equipment, levitation equipment	
	Signaling equipment	Voltage endurance
	Train control and monitoring system	
	Function and operation of train-set	Control circuit, start/stop/driver exchange, auxiliary power supply, power running, braking, air system, pantograph, protection device operation, wayside equipment interworking, air conditioning, cabin and driver equipment, door, display panel, on-board train control and monitoring system
Pre-running	Pre running	5000 km
Main line trial run	Power running, acceleration, deceleration, braking, maximum speed, running resistance, rescue operation, pantograph, inductive interference, protection device operation, noise, vibration, ride condition, wayside equipment interworking, main equipment temperature and status, emergency landing	

Table 6.15 Summary of running performance

Requirements		Test results
Acceleration	4.0 km/h/s (full load)	4.4
Jerk during acceleration	0.8 m/s^3 (full load)	0.6
Deceleration	4.0 km/h/s (full load)	4.1
Jerk during acceleration	0.8 m/s^3 (full load)	0.6
Emergency braking	4.5 km/h/s (full load)	4.8
Emergency landing with skid	Not specified km/h/s (full load)	8.2
Noise level (cabin)	70 dB(A) at 80 km/h	68
Acceleration level	0.25 g (vertical) at 80 km/h 0.175 g (lateral) at 80 km/h	0.03 0.04
Maximum speed	110 km/h	111 km/h

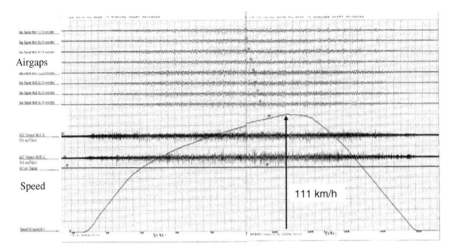

Fig. 6.51 Maximum speed test

Fig. 6.52 Test of rescue operation by drawing a disabled train

6.4.12 Operating Costs

The precise operating cost was not provided because the ECOBEE is not in
operation at this time. Only a relative comparison to wheel-on-rail light railway
vehicles was performed based on the energy consumption of ECOBEE measured
when station spacing is assumed to be 1 km. The energy consumption was measured to be 20 % higher than that of wheel vehicles. Allowing for the lower

frequency of replacement of components and the human effort involved, the total operating cost was estimated to be 60–70 % that of wheeled vehicles. The actual operating cost will be provided after the operation of ECOBEE some years in the future.

6.4.13 Deployment

Incheon International Airport and its surrounding area were selected by the Korean government as a maglev demonstration line for ECOBEE in 2007 [11]. The Incheon International Airport is a hub airport in the East Asia region that was opened in 2001 and was used by 46 million passengers in 2014, and the number of passengers that use the airport has long been rapidly increasing. Considering the increasing number of airport passengers and visitors to nearby tourist attractions, this route was chosen due to its competitiveness. The final planned route 57 km in length is a circular line traveling along the coastline of Yeongjongdo Island, where the Incheon International Airport is located (Fig. 6.53). The construction of the maglev line consists of 3 phases, the first phase being a 6.1 km stretch. The initial capital investment is 377 million USD, of which 77 million USD is dedicated to R&D activities. Construction of this system started in February 2010, and was completed in August 2012 by a consortium led by GS Engineering and Construction. All of the mandatory tests were satisfactorily conducted by mid-2014. The main features of the line are listed in Table 6.16, and some of these are summarized below.

Fig. 6.53 Bird's eye view of Incheon International Airport ECOBEE Demonstration Line and its stations

Table 6.16 Incheon International Airport ECOBEE Demonstration Line

Items	Values
Section	Airport-Yongyudong
Length	6.1 km (double track)
Stations	6 (1 island platform type and 5 side platform type)
Train sets	4
Max. operating speed	80 km/h
Design speed	110 km/h
Power line voltage	1500 VDC
Min. radius of curves	110 m
Number of switch	10 (4 for main line and 6 for depot)
Operation	ATO (driverless)

- **Passenger demand**: In 2007, due the rapidly increasing number of airport users and the many theme parks planned, the maglev line was expected to be economically feasible. Unfortunately, most of the plans for building theme parks and shopping malls were canceled because of the global currency crisis of 2008–2009, as the passenger demand was considered to be below expectations at that time. However, new hotels and other facilities are continuously being built along the line, and thus it is projected that the number of passengers will be incrementally increased. The likely potential demand for the near future is projected as 7 M people using the airport.
- **Weather conditions**: Since the line is located on an island, the crosswind has a profound effect on the lateral displacement of the electromagnet on the vehicles. When the clearance of about 15 mm is exceeded, mechanical contact between the U-shaped rail and the lateral stop may occur, resulting in impulsive accelerations of the magnets. Due to the passive nature of guidance force, the magnet guidance force capacity and bogie mechanism needed to be designed after carefully considering the crosswind effects. The interval of power rail inspection should be planned in consideration of these salty sea winds.
- **Construction costs**: The construction cost for the elevated guideway is 39 million USD/km (2009), and included countermeasures that were necessary due to region-specific conditions (poor subsoil, salt damage) and the characteristics of the route. However, in ordinary cities, the cost should not exceed 36 M$/km.

The next potential application of ECOBEE in Korea is for the city of Daejeon's Metro Line 2 [11]. With a population of 1.55 M people, Daejeon is Korea's fifth largest metropolitan city. Located in the center of Korea, Daejeon serves as a hub of transportation and is at the crossroads of major transport routes. The capital Seoul is about 50 min away by high speed train. Daejeon also serves as an administration hub with the National Government Complex. Currently, 12 national government

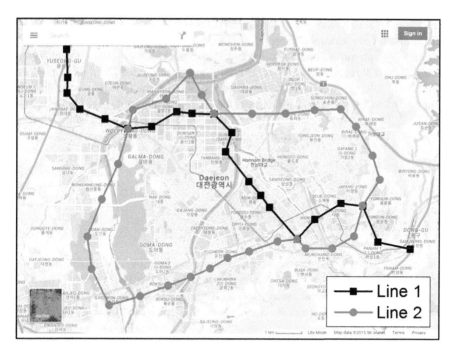

Fig. 6.54   Route of Daejeon's Metro Line 2

Table 6.17   Features of Daejeon's Metro Line 2

Items	Values
Length	36 km (underground 3 km)
Expected ridership	130,000 passengers/day
Stations	30
Total cost	1236 M$
Route	Loop type

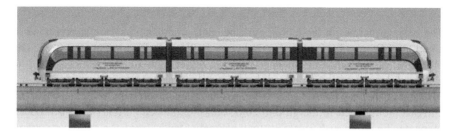

Fig. 6.55   ECOBEE in the formation of 3 cars

offices and 18 universities are located in Daejeon. Daedeok Innopolis (Daedeok Research and Development Special Zone), located in the city's northern quarter, features 28 state-run research centers and 79 private research institutes, employing many as 20,000 researchers. The route map is given in Fig. 6.54, and its features are listed in Table 6.17. It is considered likely that the trains for the line will be lengthened with 3 or more cars to meet the passenger demand (Fig. 6.55). In May 2014, ECOBEE was chosen as the vehicle type for Daejeon's Metro Line 2. However, the plan was changed in December 2014 and it was decided to use a tram.

6.5   L0

6.5.1   Introduction

On April 22 2015, the Japanese superconducting maglev system L0 (Fig. 1.4) achieved a running speed of 603 km/h, a record for any guided vehicle. The L0 is also planned by the Central Japan Railway Company to be in service over the route between Tokyo and Nagoya in 2027, as the first phase of its entry into public service. The final goal is to link Tokyo and Osaka through Nagoya with the L0, running at a maximum operating speed of 505 km/h. The estimated cost has reached JPY 91 trillion. The company is also promoting the deployment of the L0 in the USA. It is likely that the final decision regarding whether to apply L0 to the route between Washington and New York will be made in the near future. The two deployments will be discussed further later in this section. As the L0's development has taken place over more than 50 years, its details are well known in the public domain. Thus, this section outlines the main features of the levitation, propulsion and guidance technologies only.

6.5.2   Background and Features

The basic principles of levitation, guidance and propulsion are briefly presented in Chap. 4 [12]. The L0 system configuration is given in Figs. 6.56 and 6.57. The on-board superconducting magnets are on either side of the vehicle, and the 8-figured ground coils for levitation and guidance and the windings for LSM are installed inside the side wall. As explained in Chap. 4, when the vehicle moves, the currents are induced in the ground coils, and the induced currents in turn produce magnetic fields (Fig. 6.58). The two magnetic fields from the superconducting magnets and the induced currents in the ground coils generate the magnetic

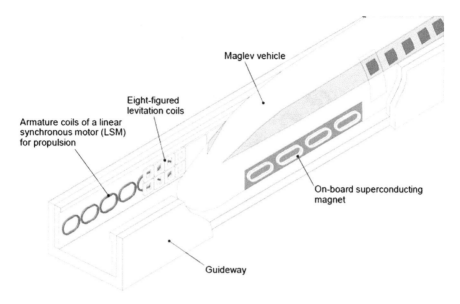

Fig. 6.56 System configuration of the L0 using superconducting magnets [13]

pressure, which provides the vehicle both with levitation and guidance forces Fig. 6.58). The propulsion method is the iron-cored long-stator LSM, wherein the superconducting magnets are also used as the field for LSM. This concept is a representative electrodynamic levitation system, with the prime advantage of having no active control requirement. The bogie in Fig. 6.57 is composed of the superconducting magnets, suspension elements, and auxiliary support wheels for landing. The guideway has a U-shape made of concrete (Fig. 6.58) [12]. Each bogie has 8 magnets, which make 2 magnetic poles on either sides. The three-phase primary windings of LSM are installed in between inner and outer layers of the side wall. The length of one section of the guideway is 12.6 m. The 8-figured null flux levitation coils (ground coils) are placed inside the primary winding. Null-flux connection is also used to link the left and right levitation coils. This connection provides the guidance forces. The coil pitch of the levitation coil is 1/3 of the superconducting coils.

Looking at the technological advancements of the L0, the following three aspects are highlighted:

- The shape of the leading car has been changed to improve the aerodynamic effects and increase the passenger capacity, with its nose now 15 m in length (Fig. 6.59).

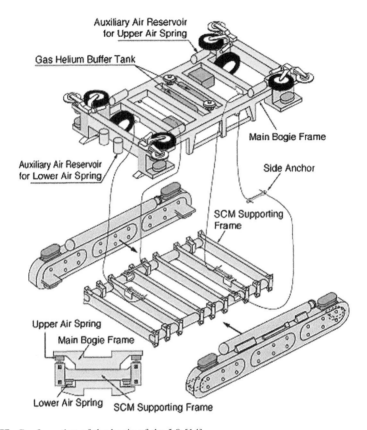

Fig. 6.57 Configuration of the bogie of the L0 [14]

- The on-board power supply from the gas turbine engine will be replaced by the inductive power collection system in the commercial version. The developer has observed that sufficient power collection is available at very high speeds, and with larger airgaps. The principle of the inductive power collection system is shown in Fig. 6.60, in which the coils for power transmission coils are installed on the guideway and the power collection coils are located beneath the vehicle.
- The third feature is the use of the high-temperature superconducting magnets in future commercial lines. If the high-temperature superconductor with higher performance and reduced cost is available, the superconducting coils would provide improved performances, such as better stability and higher current density, and a simpler cooling system. The improvements enabled by employing the high-temperature superconducting coils are illustrated in Fig. 6.61.

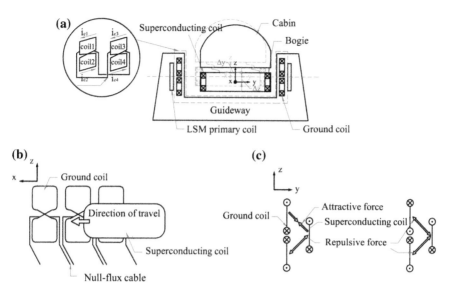

Fig. 6.58 Principles of levitation, guidance and propulsion for the L0 [15]

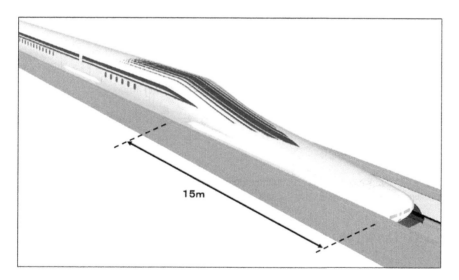

Fig. 6.59 Nose shape of the L0 system [16]

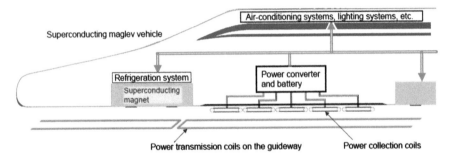

Fig. 6.60 Inductive power collection system for on-board power supply of the superconducting maglev system [13]

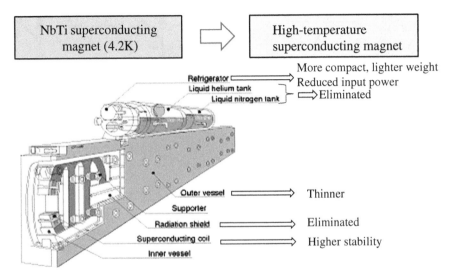

Fig. 6.61 Fundamental structure of the superconducting magnet, and advantages of a high-temperature superconducting magnet [17]

6.5.3 Deployment

Since 1997, the superconducting maglev system (including the L0 mentioned earlier) has been undergoing test runs on the Yamanashi maglev test line with the aim at achieving its commercialization, reaching a total travel distance of about 874,000 km. The test line was extended to 42.8 km in 2013, and on that line the L0 vehicle is being conducted. Ultimately, the system is planned to be applied to Chuo

Shinkansen, connecting three major metropolitan areas in Japan: Tokyo, Nagoya and Osaka. The route layout and its overview are given in Fig. 6.62 and Table 6.18. The Chuo Shinkansen project is believed to be viable.

One of the oldest planned maglev projects in the USA, the 64-km route between Washington, D.C. and Baltimore was one of two principal projects planned from 2000 to 2007 (Fig. 6.63). Following the environmental and safety studies, the private-sector team was forced to disband due to a lack of funds. However, the project has been reactivated in the past several years by the presence of the TNEM group that has reclaimed the corridor for potential use by the Japanese high-speed superconducting maglev L0. The ultimate goal is to achieve a one-hour trip time between Washington, D.C. and New York City. In addition to the TNEM group, thanks to the active promotion of the Japanese government, this is likely the most visible and active U.S. maglev program in place today.

Fig. 6.62 Cho Shinkansen operated with the superconducting maglev system between Tokyo and Osaka [13]

Table 6.18 Overview of Cho Shinkansen and Tokaido Shinkansen [13]

Items	Chuo Shinkansen (2027–)	Chuo Shinkansen (2045–)	Tokaido Shinkansen
Route	Tokyo-Nagoya	Tokyo-Osaka	Tokyo-Osaka
Length	286 km	438 km	515 km
Journey time	40 min	67 min	145 min
Max. speed	505 km/h	505 km/h	270 km/h
Total cost	5.5 trillion yen	9 trillion yen	–

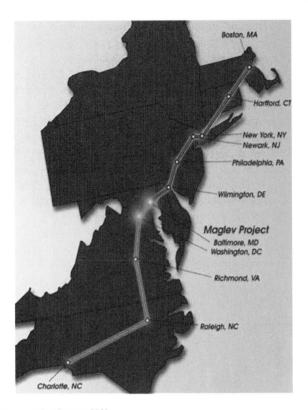

Fig. 6.63 Baltimore-Washington [18]

6.6   Chinese Urban Maglev Vehicles

6.6.1   Introduction

China is currently in an intense process of urbanization due to its rapid economic growth. Its rate of urbanization has increased from 17.9 % in 1978 to 52.6 % in 2012, and is expected to grow to 55–60 % in 2030 and to about 65 % in 2030. This very rapid urbanization means that about 10 million people in China move to cities every year. Currently, there are 13 cities in China with more than 10 million people, and 88 cities with more than 5 million people. As a result, there are serious traffic and environmental problems in Beijing, Shanghai and China's other large cities. As well, since 2000, the Chinese auto market has expanded rapidly, reaching 100 million cars. China has become the world's largest car market, and is expected to have 200 million cars by 2020. These circumstances have given rise to road congestion, air pollution and energy shortages. During the past ten years, China's large cities have been investing mainly in conventional rail to solve the traffic problems. However, due to the high construction costs of subways and the high noise level and incompatibility with surrounding buildings of the conventional light railways, the construction of conventional rail transportation has become difficult. To overcome these problems, the maglev train has emerged as an alternative to the conventional systems, as the maglev system has environmental, operational and economic benefits compared to conventional rail. In November 2004, Zhou Ganzhi published a thesis on "the basic situation of the low-speed maglev train in Nagoya" which introduced the technical and economic characteristics of the low speed maglev system and suggested that the application of low speed maglev technology in the field of urban transit in China be studied. In January 2005, he also suggested that the Shanghai city leaders and the Shanghai maglev research center organize research on low speed maglev transportation engineering. Subsequently, the Shanghai Maglev Research Center, with support from the Shanghai municipal government, began the design and construction of a low speed Maglev test line. Since then, from 2005 to 2011, three R&D groups have built three 1.5–1.7 km long test tracks in Shanghai, Tangshan (Hebei province) and Zhuzhou (Hunan province).

6.6.2 Status of Vehicle Development

For urban maglev vehicles, three models have been in development, on 3 test tracks [19].

- **Shanghai**: From May 2005 to the end of 2007, Shanghai Maglev Transportation Engineering R&D Center in association with Shanghai Electric Group built a three-section urban maglev train and urban maglev test line (Fig. 6.64). The test track is composed of a 1704 m main track, a 276 m depot line and a set of two-way switches. The track gauge is 1.9 m. The three-car train (47 m in length and 2.8 m in width) was constructed to have a maximum design speed of 100 km/h. The vehicle uses the electromagnetic levitation system (EMS) and the short-stator linear induction motor for propulsion. The smallest lateral curve radius of the test track is 50 m, the smallest vertical curve radius 1500 m and the maximum gradient 70 ‰. The vehicle attained a maximum speed of 101 km/h in April 2008. During the performance tests, the environmental noise levels, stray magnetic fields, suspension stability and energy consumption were verified to satisfy the requirements for urban transit. This system is continuing to pursue maglev R&D and its application.
- **Tangshan**: Beijing Holding Maglev Technology Development Co., Ltd, National Defense Science and Technology University and Tangshan Railway Vehicle Co., Ltd built the 1.5 km Low-speed Maglev Engineering Test Track in Tangshan and put it into trial operation in 2008 (Fig. 6.65). Compared to the

Fig. 6.64 Shanghai urban
(low speed) maglev test track
[19]

Shanghai Maglev Test track, the main difference is the track gauge of 2000 mm
and the 3000 mm width of the vehicle. The vehicle specifications are listed in
Table 6.19. The distance covered was more than 70,000 km, and 110 tests to
evaluate the safety and availability were completed.

Table 6.19 Tangshan Maglev vehicle specifications

Items	Characteristics	Remarks
Vehicle length (m)	15/14	First/middle car
Vehicle height (m)	3.83	
Vehicle width (m)	3	
No. of levitation modules	10	
Airgap (mm)	8	
Empty load (ton)	23/22	First/middle car
Full load (ton)	33/34	First/middle car
Passenger capacity (person)	132/168	6 persons/m^2
Power supply	DC 1500 V/DC 750 V	At side bottom of the third and fourth rails
Power dissipation of magnet (kW/ton)	0.7	
LIM power (kVA)	1100	
Running speed (km/h)	100	
Construction speed (km/h)	150	
Noise (dB(A))	64	Beyond 10 m
Acceleration (m/s^2)	1.1	
Deceleration (m/s^2)	1.1	
Emergency deceleration (m/s^2)	1.3	
Train control	MTCS	
Running control	MATC (ATP, ATS, ATO)	

Fig. 6.65 Tangshan urban maglev test track [19]

Fig. 6.66 Zhuzhu maglev test track [20]

- **Zhuzhou**: The Zhuzhou Electric Locomotive Ltd, Southwest Jiaotong University and China Railway Eryuan Engineering Group Co., Ltd built the 1.5 km Test Track in Zhuzhou for the Low-speed Maglev vehicle, and put it into trial operation in early 2012 (Fig. 6.66). The 3-section train with a width of 2800 mm and a track gauge of 1860 mm was constructed. As of June 2014, the train had covered a total distance of 16,000 km. The main performance and availability evaluations were completed.

6.6.3 Deployment

In China, two urban maglev lines are currently under construction, and are scheduled to open in 2016 [19].

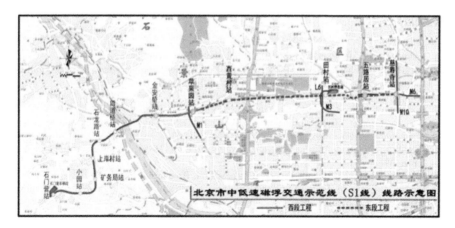

Fig. 6.67 Beijing S1 maglev line [21]

- **Beijing S1**: The construction of the urban maglev line S1 in Beijing started in October 2013, and it is planned to complete the test runs in December 2015. The total length of the S1 is 10.2 km, with 8 stations (Fig. 6.67).
- **Changsha**: Changsha Maglev Line connects the airport and the south station of the high-speed railway. The line is completely an elevated guideway with a length of 18.5 km and three stations (Fig. 6.68). The train set consists of 3 cars. This project officially started on May 16th 2014, and the completion of the test runs is expected by the end of 2015.

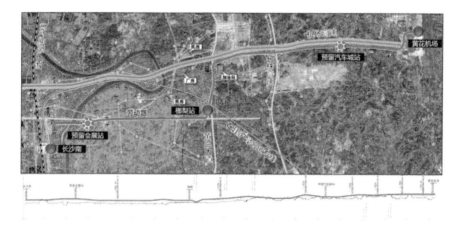

Fig. 6.68 Changsha maglev line [22]

References

1. Image courtesy of ThyssenKrupp Transrapid GmbH/StoiberProductions
2. Becker P, Frantzheld J, Loeser F, Zheng Q (2014) Transrapid-proven solution meeting current and future transport needs. Maglev 2014, Rio de Janeiro, Brazil
3. Wolters C (2008) Latest generation maglev vehicle TR09. Maglev 2008. San Diego, USA
4. Courtesy of Aichi Kosoku Kotsu. http://www.linimo.jp
5. U.S. Department of Transportation (2004) Chubu HSST maglev system evaluation and adaptability for US urban Maglev, FTA-MD-26-7029-03.8
6. Yasuda Y, Fujino M, Tanaka M, Syunzo I (2004) The first HSST maglev commercial train in Japan, Maglev 2004. Shanghai, China
7. Aichi Rapid Transit. Linimo Technical Information, Brochure
8. Morita M, Iwaya M, Fujino M (2006) The characteristics of the levitation system of Linimo (HSST system), Maglev 2006. Dresden, Germany
9. Yuyama Y (2004) The ToBu Kyuryo Line (popular name: Linimo) a magnetic levitation system. Maglev 2004. Shanghai, China
10. Park DY, Shin BC, Han H (2009) Korea's urban maglev program. Proc IEEE 97(11): 1886–1891
11. Shin BC, Park DY, Han HS, Lee JM, Baik SH, Beak JG, Kang HS (2014) Korea's first urban maglev system. Maglev 2014, Rio de Janeiro, Brazil
12. Ota S, Yoshioka H, Murai T, Terumichi Y (2014) Fundamental study on preview vibration control for the superconducting maglev. Maglev 2014, Rio de Janeiro, Brazil
13. Ohsaki H (2014). Japanese superconducting maglev-development and commercial service plan. Maglev 2014, Rio de Janeiro, Brazil
14. Source: http://www.rtri.or.jp/rd/division/rd79/yamanashi/english/html/mlu002n_E.html#bogie
15. Ohashi S, Ohsaki H, Masada E (2000) Running characteristics of the superconducting magnetically levitated train in the case of superconducting coil quenching. Electr Eng Jpn 130(1):95–105
16. Source: http://jr-central.co.jp/news/release/_pdf/000009381.pdf
17. Source: http://www.rtri.or.jp/rd/division/rd79/yamanashi/english/html/yamanashi_scm_E.html
18. Blow LE (2014) Status of maglev projects in the USA. Maglev 2014, Rio de Janeiro, Brazil
19. Lin G, Sheng X (2014) Application and development of maglev transportation in China. Maglev 2014, Rio de Janeiro, Brazil
20. Photo courtesy of CSR Zhuzhu Electric Locomotive Co., Ltd
21. Image courtesy of Beijing Holding maglev Technology Development Co., Ltd
22. Image courtesy of China Railway Eryuan Engineering Group Co., Ltd

Chapter 7
Applications

7.1 Introduction

Thanks to the technically successful implementation of magnetic levitation technology in trains and the increasing demand for contactless operation in various areas, there have recently been some attempts to apply magnetic levitation technology in areas where contactless operation is a prime design consideration. Moreover, the use of the subsystems utilized in the maglev systems, i.e. linear motors or magnets, is also being proposed, particularly in areas such as transportation, defense and aerospace, among others. This chapter introduces such newly proposed applications with the aim of highlighting the use of magnetic levitation in those applications. The chapter may help the reader imagine their own new applications using magnetic levitation technology.

7.2 Clean Conveyor

Most conveyors in automation are based mainly on mechanical elements, such as rollers. Particularly in the areas of semiconductor and display, like LCD industries, extreme cleanness is of importance for the economic mass production of wider displays, highlighting the need for a new generation conveyor with nearly zero particles. For this reason, the contact-free magnetic conveyor is attracting attention in such industries. A contact-free magnetic conveyor can have two types of configurations. The first design is the active type, in which the levitated moving object has its own power, though supplied by an external power source, to lift itself, the object having the lift magnets. On the other hand, in the second design, the moving object does not have the lift magnets, as the magnets are placed on fixed objects. These two types of configurations are briefly introduced in the following sections.

© Springer Science+Business Media Dordrecht 2016
H.-S. Han and D.-S. Kim, *Magnetic Levitation*,
Springer Tracts on Transportation and Traffic 13,
DOI 10.1007/978-94-017-7524-3_7

7.2.1 Active Type

As LCD TVs become wider and larger, extremely clean conveyors are required in order to reduce the defects that can be caused by the adhesion of undesired particles during the manufacturing process. Most conventional roller conveyors are limited in terms of their capacity to protect against the particles generated by the many friction elements. In addition, the need for conveyors that are capable of higher speeds is increasing. To solve these two problems, the magnetic levitation method is emerging as an alternative to roller conveyors, and offers both extreme cleanness and higher speed operations. In an active type system, the vehicle carrying loads such as LCDs has levitators on it. There has been an attempt to apply the U-shaped hybrid magnets and LIM to a vehicle carrying LCD Glasses [1, 2]. The total weight of the conveyor is 760 kg, including a payload of 350 kg. The system performance requirements are listed in Table 7.1. The nominal airgap is 3 mm and operating speed is 2 m/s, with a payload of 350 kg [2].

- **Configuration**: The general view of the magnetic conveyor and its concept are given in Figs. 7.1 and 7.2, respectively. The system configuration is almost the same as for the ECOBEE, with the exception of the hybrid magnets and the contactless power supply. As such, because it has fully non-contact operation, the system generates no particles when operating at high speeds.
- **Lift magnet**: Hybrid magnets are used for levitation to decrease the required power consumption and increase the airgap. Eight magnets are used, with each corner having a magnet set. The design specifications are listed in Table 7.2, and the magnet design specifications are given in Fig. 7.3. The levitation-current characteristics are shown in Fig. 7.4, which shows that the required lift force is derived at 0 A. This indicates that the nominal lift force is provided only by the permanent magnet in the hybrid magnet. Conceptually, the currents are used only for stabilization of dynamic force (deviation from nominal value), resulting in improvements in the lift force/input power ratio. From an operational perspective, a technique to protect against the contact of the magnet and the rail should be considered.

Table 7.1 System performance requirements

Item	Value
Payload	350 kg
Nominal airgap	3 mm
Max. airgap	5 mm
Airgap deviation	0.3 mm
Acceleration	1 m/s^2
Max. speed	3 m/s
Operating speed	2 m/s

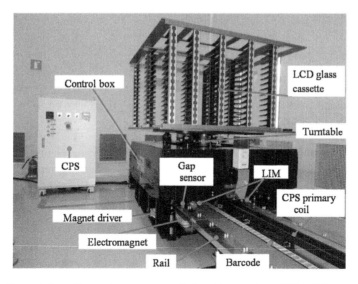

Fig. 7.1 General view of a maglev conveyor with hybrid magnets and LIMs [2]

Fig. 7.2 Configuration of the
maglev conveyor

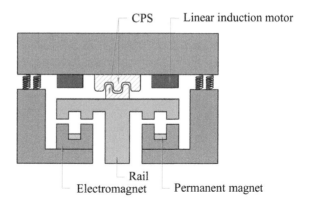

- **Guidance**: Because of the use of U-shaped magnets, the guidance forces are basically provided. However, a staggered configuration is used to reduce yaw motion of the vehicle while the turntable rotates (Fig. 7.5).
- **Thrust**: For propulsion, two LIMs are used on each side. The design of LIM follows the typical design procedure. The design specifications of the LIM are listed in Table 7.3. One inverter controls two LIMs.
- **Power supply**: The type of contactless power supply for the vehicle is an inductive one, where the primary coils are installed on the rail and the two current pick-ups are on-board. The power capacity of the inverter for the primary coils is 20 kW, while the power source for levitation and system control is 8 kW with 150 V, and 5 kW with 300 V for LIM.

Table 7.2 Hybrid magnet design specifications

Item	Unit	Value	Item	Unit	Value
Pole length	mm	200	Core material		Pure iron
Pole width	mm	20	PM material		NdFeB (N40SH)
Pole height	mm	60	Coil material		Copper
Yoke thickness	mm	20	Coil mass	kg	5.82
Yoke length	mm	60	Core mass	kg	3.55
Yoke width	mm	40	Total magnet mass	kg	9.37
Coil turns	turn	360	Nominal lift force	kgf	760.9
Coil height	mm	38	Ratio of levitation force/magnet weight		10
Coil width	mm	18	Total Ampere turn	kAT	7.2
Coil section area	mm^2	1.13	Nominal current	A	0
Max. current	A	10			

Fig. 7.3 Configuration of the hybrid magnet used

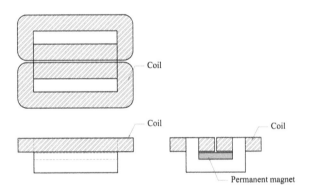

Fig. 7.4 Levitation force-current at different airgaps

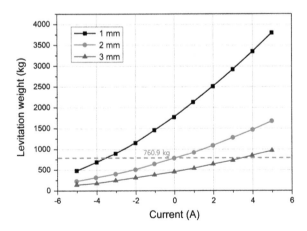

Fig. 7.5 Staggered magnets

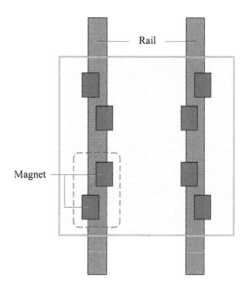

Table 7.3 LIM specifications

Item	Specification	Unit
Primary		
Input voltage	220	V
Input current	10	A
Thrust	323	N
Normal force	640	N
Phases	3	Phase
Poles	8	Pole
Secondary		
Material	Aluminum	–
Al thickness	2.5	mm
Overhang length	20	mm
Back iron thickness	12.5	mm

- **Position detection**: The position detection needed for LIM control is realized using the barcode system available in the room environment.
- **System and levitation control**: The system architecture for system control is given in Fig. 7.6. The basic configuration is conceptually similar to that of a maglev train. The levitation controller is based on the conventional phase lead-lag compensator in continuous time domain, and is discretized using the bilinear transform (also known as Tustin's method). In addition, a zero-power control scheme is applied to balance the attraction force and the total weight. Some experiments were performed to verify the designed control method. Figure 7.7 shows that the performance requirements of a nominal airgap of

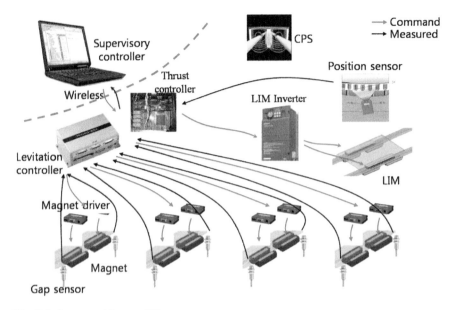

Fig. 7.6 System architecture [2]

3 mm with a deviation ±0.3 mm are satisfied. Here, it can be noted that at steady state, the current is not exactly 0 A, though low. In practice, a certain bias current is needed for stable operation.

7.2.2 Passive Type

For the OLED TV manufacturing process, some special constraints are imposed on transferring the glass. The constraints are vacuum environment, zero-particle and dead section of any actuator. This implies that the tray carrying glass should not have its own power for suspension, guidance or propulsion, as any power transmission method for the tray may generate undesired particles and gas. To meet those requirements, a so-called passive type maglev conveyor has been proposed using the electromagnets and LSM [3–6]. The configuration of the magnetic conveyor is presented in Figs. 7.8 and 7.9. The magnets fixed to the structure attract the tray upwards and guide it laterally. The magnets are placed at an equal distance all along the track, requiring a number of magnets. Then, the tray is propelled by the external LSM. Consequently, the tray without power can be moved in a vacuum chamber, generating no particles. Some technical features are summarized as follows.

Fig. 7.7 Levitation control characteristics of the maglev conveyor

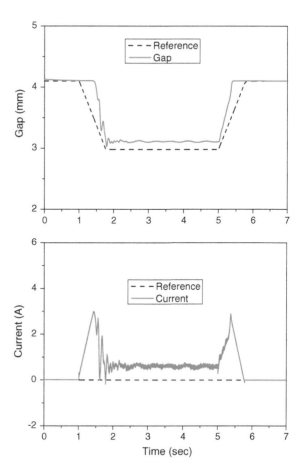

- A number of magnets are needed for the smooth movement of the tray. As can be expected, the magnets and their associated drivers and sensors are installed discretely along the whole length of the chamber. As a result, this system requires a high number of magnets compared to the active type conveyor. Due to the need for many magnets, there are still many aspects open for research and development.
- Magnet spacing is the key design variable in this configuration. The larger the spacing is, the more unsteady motion occurs. This is because of the discontinuous or impulsive application of attraction forces to the tray. Figure 7.10 shows the behavior. The magnet entering the tray attracts it; in contrast, the magnet exiting drops it. These force actions result in the pitch motion of the tray. To reduce the pitch motion, the magnet spacing and force applications must be optimized.

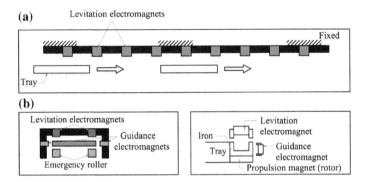

(a)

(b)

Fig. 7.8 Configuration of the passive tray conveyor system: **a** side view, **b** front view

Fig. 7.9 General view of the passive tray conveyor system

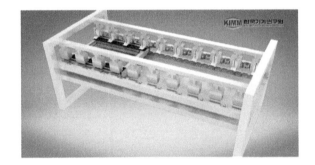

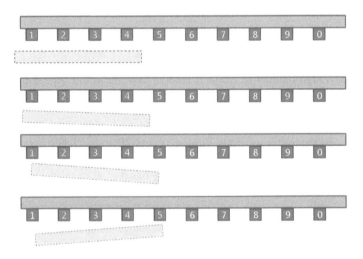

Fig. 7.10 Magnets switching and the expected tray behavior

- In practice, some vibrations may occur due to the flexibility of the structure fixing the magnets. Heavier and stiffer structures are desirable to reduce the vibrations, resulting in a compromise between cost and vibration reduction.
- **Levitation control scheme**: Unlike the active type using accelerometers in levitation control, since this passive type probably can't place any accelerometers on the tray, the control scheme has to be based on external airgap sensors only. A control law that uses airgap (y), its velocity $(\dot{y})$ and acceleration $\left(\ddot{y}\right)$ was used by Kim [3]. To derive the airgap velocity and acceleration, the numerical differentiation formula described in Chap. 2 may be employed. The control voltage is determined by

$$\Delta v = k_p \times (y - reference) + k_v \times \dot{y} + k_a \times \ddot{y} \tag{7.1}$$

where k_p, k_v and k_a represent the control gains for the airgap, its velocity and acceleration, respectively. Though each magnet is basically controlled independently by Eq. (7.1), a special scheme, a feed forward term, to reduce the pitch motion of the tray was added to the feedback loop Eq. (7.1). The function of the feed forward is to apply the predetermined attraction forces to magnets. The predetermined forces have the form of a triangle, as shown in Fig. 7.11, making the moment on the tray always zero regardless of the longitudinal position of the tray. To provide the predetermined force with the position of the magnet, the bias current is first calculated with the position and then added to the control voltage Eq. (7.1) through the relationship between voltage and current. With this scheme, a reduction in the pitch motion of the tray can be expected.

- **Guidance control**: Lateral clearance between the tray and frame is controlled using the same scheme described in Sect. 5.6.
- **Propulsion**: The thrust is provided by the external LSM whose primary is installed on the frame and whose secondary field of permanent magnets attached underneath the tray. A barcode reader is used to detect the position of the tray, the measured position being used both in the LSM and in determining the feed forward term.

This passive type conveyor system has potential uses in other conveyors needing cleanliness. However, some outstanding problems remain, particularly the need for cost reduction due the high number of magnets used and the vacuum operation environment.

Fig. 7.11 Levitation force
control scheme to make the
moment zero

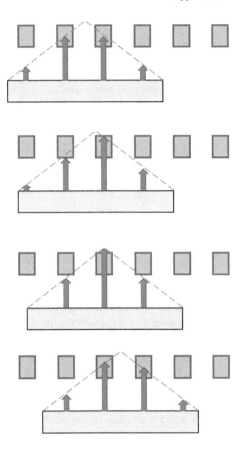

7.3 Ropeless Elevator

In modern skyscrapers and very large buildings, conventional rope type elevators
face a variety of problems related to speed, height, space, passenger capacity, etc.
To overcome such limitations, a ropeless elevator system with a linear drive and an
electromagnetic linear guide has recently been in development [7–9]. The basic
concept is to replace the contact shoe guiding the cabin with magnetic levitation
(magnetic linear guiding), and to replace the steel rope and counter weight with
LSM (Fig. 7.12) [7]. This ropeless elevator could offer more comfortable transportation, higher speeds and improved operational flexibility. For instance, since
multiple cabins can run on the same shaft as well as move horizontally, the space
required for the elevator system can be reduced, increasing the potential profit to the
building owner by providing him/her with more space to rent. The CPS(Contactless
Power Supply) is employed to transmit the required energy to the cabin [9]. As a
result, the cabin can run without any physical contact.

Fig. 7.12 Configuration of ropeless elevator with LSM and magnetic linear guide

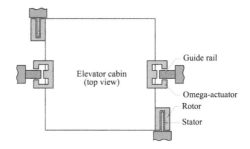

Guide rail

Elevator cabin (top view)

Omega-actuator

Rotor

Stator

Fig. 7.13 Configuration of linear guide using hybrid magnet and steel rail

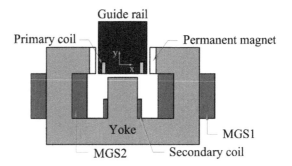

Guide rail

Primary coil

Permanent magnet

Yoke

MGS1

MGS2

Secondary coil

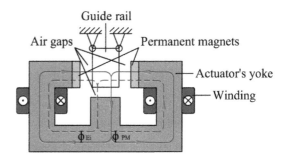

Guide rail

Air gaps

Permanent magnets

Actuator's yoke

Winding

- **Linear guide**: Four hybrid magnets, which are called an Omega actuator, are used to maintain the clearance between the guiderail and the Omega actuator (Fig. 7.13) [7–9]. Two magnets are on top of the elevator roof and two are under the floor. The magnetic flux path from the magnet through the guiderail generates an attraction force between them. The deviations in airgap are stabilized by adjusting currents in the magnet coils (MGS1, MGS2). Eddy-current sensors are used to measure airgaps in the directions of the X and Y axes. A DOF control scheme is employed to stabilize the cabin in its nominal position at the center of the shaft. With four actuators, the 6 DOF of the cabin can be controlled. Using the bias permanent magnetic excitation in the actuator yokes, electric energy is only required for correction of the nominal position. The total power consumption of the guiding system and the on-board electronics is less

Fig. 7.14 LSM with inclined
permanent magnets as rotor
[7]

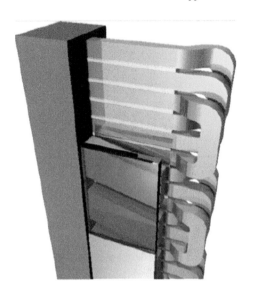

than 200 W. The cabin is propelled by two LSMs (Fig. 7.14) [7]. The coils on
the wall and the permanent magnets on the cabin form the LSM. To reduce force
ripples, the permanent magnets used as a rotor were skewed. The regenerative
braking by the LSM reduces the energy consumption. The primary coils inserted
into the guiderail and the secondary coils in the Omega actuator form the CPS.
This new generation of ropeless elevator offers the following advantages and
disadvantages compared to conventional rope type elevator.

- **Advantages**: More rentable space, reduced average ride time, and higher
  transport capacity
- **Disadvantages**: More complex electronics and mechanics, higher power
  requirement, and high investment costs

An economic analysis is first required to determine whether this ropeless system
should be constructed.

7.4 Amusement Rides

7.4.1 Hover Board

The idea of a board that a person can ride on while hovering over the ground has
been a staple of science fiction movies for decades. Interestingly, such hover boards
for rides have been recently presented for sale. Magnetic levitation can also be used
to form such hover boards. There are two types of magnetic levitation methods that
can be used for a hover board. The first type is the active type magnetic hover board
having magnetic wheels itself for levitation, with the ground being formed simply

Fig. 7.15 Conceptual illustration of an active type magnetic hover board having 4 magnetic wheels

Fig. 7.16 General view of an active type magnetic hover board having 2 magnetic wheels [13]

of aluminum sheets [10]. The second type is a passive type magnetic hover board without magnetic wheels, in which the magnetic wheels are installed on the ground [11]. Figure 7.15 shows the active type magnetic hover patented in 2012 [10]. The principle of its levitation was described in Sect. 3.5 [12]. With on-board magnetic wheels with permanent magnets, the board can be levitated over aluminum sheets fixed on the ground. Self-propulsion is theoretically possible by tilting the magnetic wheels, i.e. by using magnetic drag forces. A company released a hover board of this type in 2014 (Fig. 7.16) [13]. In practice, in such a type of hover hoard, power supply would be a problem because the lift force/input power ratio is low compared to in the attraction type. The passive type magnetic hover board is conceptually shown in Fig. 7.17 [11]. In this design, the magnetic wheels are installed on the ground, the board being just an aluminum plate. This means that a high number of magnetic wheels needs to be placed on the ground, increasing the cost. In addition, the temperature rise due to induced eddy currents in the board may be a critical problem. Of course, the fatal limitation of these hover boards is that the board can only hover above conductive sheets, not regular ground.

Fig. 7.17 Conceptual illustration of a passive-type magnetic hover board

7.4.2 Hover Car

As an extension of the hover board above, magnetic hover cars have been conceptually proposed by several inventors [14]. The magnetic wheel technology given in Sect. 3.5 can be one method for the realization of this invention. To drive and steer the car, linear motors in two directions can be employed. The basic principles of levitation and propulsion are the same as those in Sect. 3.5. Two imaginations of magnetic hover cars are illustrated in Figs. 7.18 and 7.19. The features of these cars are the same as those of the magnetic hover board. These magnetic hover cars also can run over a conductive plate only.

Fig. 7.18 Conceptual illustration of a magnetic hover car for amusement rides

Fig. 7.19 Conceptual
illustration of a magnetic
hover car for carrying
passengers

Fig. 7.19 Conceptual illustration of a magnetic hover car for carrying passengers

7.5 Cargo Conveyor

There is also a proposition that magnetic levitation and linear drive technologies could be applied to cargo transportation (Fig. 7.20). The motivation of the application of such technologies to cargo transportation is similar to that of passenger transportation. The expected benefit is a new transportation solution that can meet the urgent need for cleaner, more efficient means of goods movement. This system is based on the Halbach array permanents and LSM. The details of these were described in Chap. 3.

Fig. 7.20 Maglev cargo conveyor concept [14]

7.6   Passenger Transport

7.6.1   Personal Rapid Transit

PRT(Personal Rapid Transit) is best characterized as a guided taxi. As it runs on a purpose-built guideway, it is not affected by the road traffic (Fig. 7.21). This PRT is suspended and propelled by electromagnets and LIMs, respectively. It is claimed by the skyTran company that this concept is capable of carrying passengers in a fast, safe, green, and economical manner.

7.6.2   Vacuum Tube Car

In ground transportation at very high speeds, speed is limited by air resistance. To reduce air resistance, partly evacuated tubes or tunnels could be used with a guideway. A vacuum (evacuated) tube car is a proposed design for very-high-speed guided transportation. The lack of air resistance could permit vacuum tube cars to travel at very high speeds. The most promising technologies to realize these vacuum tube cars are maglev and linear motor technologies. This car is expected to offer a silent, low cost, safe form of transportation that is faster than jets. Several propositions based on this technology have been made since the early 1960s. Two of them are conceptually and experimentally shown in Fig. 7.22 [16] and Fig. 7.23 [17].

Fig. 7.21 Maglev personal
rapid transit [15]

Fig. 7.22 Concept of a
vacuum tube car [16]

Fig. 7.23 Small-scale
vacuum tube vehicle [17]

Fig. 7.24 A track on test
model scale for lower velocity
magnetic launch assist [18]

Fig. 7.25 A prior concept for
a horizontal launch assist
system based on
superconducting maglev tech,
but at far lesser velocity:
MagLifter [19]

7.7 Aerospace Systems

7.7.1 Space Launch

To truly open the space era, the cost of transporting passengers and cargo from the
ground to low earth orbit must be dramatically reduced. The StarTram project aims
to reduce the cost per kg for cargo from approximately $10,000 to just $50.
StarTram is a proposal for a maglev space launch system, which would require
neither rockets nor propellant to launch a payload into space. The basic concept is
to use a sled carrying a space craft that is levitated by superconducting magnets and
propelled by LSM at ultra-high speeds. Depending on the purposes, the system
configurations can vary. Figures 7.24 and 7.25 are examples of launch assist.
Figure 7.26 shows the concept for launching a space craft to an altitude of 22 km
above sea level. Though these concepts need various extremely advanced technologies, as can be expected, the maglev and linear drive form the basis for their
realization.

7.7.2 Flight Testing

As an alternative to flight tests, the maglev track is used to provide a ground-based
test environment that can accurately simulate the vibrations of a real flight test at a
maximum attained speed of 800 km/h (Fig. 7.27). The superconducting magnets are
used to lift the sled carrying the missile in the guideway. The levitation principle is
the same as that described in Chap. 4. The magnetic forces act on the sled as the
aerodynamic loads seen during actual flight. The smoothness of the ride due to the
magnetic cushion is the maglev's key asset, and enables the preservation of the

Fig. 7.26 StarTram Generation 2 launching a spacecraft, a megastructure with an elevated launch tube 22 km above sea level (a bit less above local terrain), a more ambitious proposal than the surface-based Gen-1 and Gen-1.5 versions [20]

most delicate components throughout the duration of the test. Consequently, a complete record of the event can be successfully captured for further analysis. The Air Force Research Laboratory claimed that maglev and traditional steel rail can acquire roughly 90 % of the data needed at about 10 % of the cost (Fig. 7.27).

7.7.3 Space Elevator

The space elevator was first conceived in 1895 by the Russian scientist Konstantin Tsiolkovsky as a cable fixed to the equator and reaching into space (Figs. 7.28 and 7.29). A counterweight at the upper end keeps the center of mass well above geostationary orbit level at the altitude of 35,790 km. This produces enough upward centrifugal force from the Earth's rotation to fully counter the downward gravity, keeping the cable upright and taut. Climbers carry cargo up and down the cable. Maglev and linear drive are used to operate the climbers along the cable. As can be

Fig. 7.27 Superconducting maglev track for flight tests [21]

Fig. 7.28 Principle of the
space elevator

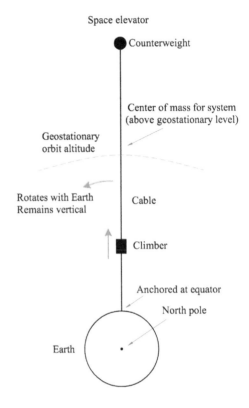

imagined, the climber can be guided without contact by electromagnets, similar to
the ropeless elevator. And with LSM, it can move upwards and downwards at very
high speeds.

Fig. 7.29 Artist's concept:
space elevator [22]

References

1. Kim K-J, Han H-S, Kim C-H, Yang S-J (2013) Dynamic analysis of a Maglev conveyor using an EM-PM hybrid magnet. J Electr Eng Technol 8(6):1571–1578
2. Kim C-H, Lee J-M, Han H-S, Lee C-W (2011) Development of a maglev LCD glass conveyor, Maglev 2011, Daejeon, Korea
3. Kim C-H, Lim J, Ahn C, Park J, Park DY (2013) Control design of passive magnetic levitation tray. In: 2013 International conference on electrical machines and systems, Busan, Korea
4. Park J-W, Kim C-H, Park DY, Ahn Changsun (2014) Controller design with high fidelity model for a passive maglev tray system. Int J Precision Eng Manufact 15(8):1521–1528
5. Li SE, Park J-W, Ahn C (2015) Design and control of a passive magnetic levitation carrier system. Int J Precis Eng Manuf 16(4):693–700
6. Kim B-S, Park J-K, Kim D-I, Kim S-M, Choi H-G (2014) Integrated dynamic simulation of a magnetic bearing stage compatible for particle free environment. In: Proceedings of the 14th euspen international conference, Dubrovnik, Croatia
7. Appunn R, Hameyer K (2014) Modern high speed elevator systems for skyscrapers, Maglev 2014, Rio de Janeiro, Brazil
8. Appunn R, Schmulling B, Hameyer K (2010) Electromagnetic guiding of vertical transportation vehicles: experimental evaluation. IEEE Trans Ind Electron 57(1):335–343
9. Appunn R, Hameyer K (2014) Contactless power supply for magnetically levitated elevator systems using a SMC hybrid actuator. In: 2014 International Conference on Electrical Machines, Berlin, Germany
10. Patent: KR 1012156300000 (2012) Magnetic levitation system having Halbach array
11. Patent: KR 1011740920000 (2012) Magnetic levitation system having Halbach array
12. Ogawa K, Horiuchi Y, Fujii N (1997) Calculation of electromagnetic forces for magnet wheels. IEEE Trans Magn 33(2):2069–2072
13. Photo courtesy of Arx Pax. http://arxpax.com/
14. Image courtesy of General Atomics. http://www.ga.com/ecco/
15. Image courtesy of skyTran, Inc. http://www.skytran.us/images/
16. Image courtesy of ET3 Global Alliance, Inc. http://et3.com/
17. Photo courtesy of Southwest Jiaotong University
18. Source: http://mix.msfc.nasa.gov/IMAGES/HIGH/9906088.jpg
19. Source: http://www.msfc.nasa.gov/NEWSROOM/news/photos/images/rbcc_maglifter.jpg
20. Source: http://science.ksc.nasa.gov/shuttle/nexgen/Misc_No_Link_to_Mains/Maglev/Renderings/Star%20Tram%20launch1.jpg
21. Source: http://www.holloman.af.mil/news/story.asp?id=123336086
22. Source: http://nix.larc.nasa.gov/info;jsessionid=b5c6c7nhf61hi?id=MSFC-0302060&orgid=11

Index

Transfer function, 26, 27, 122, 124, 126,
   127, 152
Transrapid, 3, 4, 11, 75, 85, 86, 167, 168,
   171, 172

U
U-core magnet, 91–93, 96, 104, 146, 148, 151
Upper critical field, 53

V
Vacuum tube car, 238, 239

Vacuum tube train, 238
Vehicle/guideway interaction, 157, 161
Virtual prototyping model, 159, 160

W
Window area, 92, 93, 95

Y
Yaw control, 155, 156

CPSIA information can be obtained
at www.ICGtesting.com
Printed in the USA
LVOW05*1029080617

537397LV00002B/3/P